CENTER FOR ORIGINS RESEARCH

# Issues
### IN CREATION

# Animal and Plant Baramins

Todd Charles Wood
Center for Origins Research

Center for Origins Research Issues in Creation
Number 3
November 7, 2008

WIPF & STOCK · Eugene, Oregon

ANIMAL AND PLANT BARAMINS

Copyright © 2008 Center for Origins Research. All rights reserved. Except for brief quotations in critical publications or reviews, no part of this book may be reproduced in any manner without prior written permission from the publisher. Write: Permissions, Wipf and Stock Publishers, 199 W. 8th Ave., Eugene, OR 97401.

ISBN 13: 978-1-60608-325-2

www.wipfandstock.com

Manufactured in the U.S.A.

## Abstract

To increase the number of identified holobaramins, 63 character sets from 61 different groups of animals and plants were examined using baraminic distance correlation and multidimensional scaling. Forty holobaramins and twelve monobaramins were identified. With previously published work, these new groups bring the totals to 49 holobaramins and 57 monobaramins. Newly identified holobaramins include the first arthropod, fern, and annelid holobaramins. The species counts for the 49 holobaramins are log normally distributed, and a complex model of post-Flood diversification that includes exponential growth, exponential changes in carrying capacity, and exponential decay in diversification rate is proposed to account for this. Despite the increased size of the database of holobaramins, it remains unclear what kind of characters should be used to recognize holobaramins. Several of the newly-identified holobaramins will have important implications in other areas of creationist biology, including natural evil and diversification.

# Contents

1. **Introduction**    1
   1.1. Baraminology    1
   1.2. Methods    1
   1.3. Key to Abbreviations    4

2. **Animalia**    5
   2.1. Galagonidae    5
   2.2. Dasypodidae    8
   2.3. Leporidae    11
   2.4. Felidae    14
   2.5. Viverridae    20
   2.6. Phocidae    24
   2.7. Erinaceidae    28
   2.8. Talpidae    32
   2.9. Tenrecidae    35
   2.10. Mormoopidae    39
   2.11. Phyllostomidae    42
   2.12. Hippopotamidae    46
   2.13. Brontotheriidae    50
   2.14. Rhinocerotidae    53
   2.15. Anserinae    57
   2.16. Ardeidae    61
   2.17. Falconidae    65
   2.18. Alcidae    68
   2.19. Fringillidae    71
   2.20. Pipridae    78
   2.21. Spheniscidae    81
   2.22. Pygopodidae    85
   2.23. Salamandridae    88
   2.24. Gadidae    91
   2.25. Liparidae    95
   2.26. Gasterosteidae    99
   2.27. Stomiidae    102
   2.28. Epimeriidae and Iphimediidae    105
   2.29. Pholcidae    108
   2.30. Theridiidae    112
   2.31. Sarcoptidae    116
   2.32. Ixodidae    120

| | |
|---|---|
| 2.33. Sironidae | 124 |
| 2.34. Bothriuridae and other Scorpionoidea | 127 |
| 2.35. Histeridae | 131 |
| 2.36. Coelopidae | 134 |
| 2.37. Lophopidae | 137 |
| 2.38. Membracidae | 141 |
| 2.39. Phyllodocidae | 146 |
| **3. Plantae** | **149** |
| 3.1. Saururaceae | 149 |
| 3.2. Aristolochiaceae | 152 |
| 3.3. Nymphaeaceae | 155 |
| 3.4. Moringaceae | 158 |
| 3.5. Alseuosmiaceae | 161 |
| 3.6. Cunoniaceae | 164 |
| 3.7. Robinieae | 168 |
| 3.8. Olacaceae | 171 |
| 3.9. Celastraceae | 175 |
| 3.10. Rubiaceae | 179 |
| 3.11. Zosteraceae | 182 |
| 3.12. Lemnaceae | 185 |
| 3.13. Commelinaceae | 188 |
| 3.14. Rapateaceae | 192 |
| 3.15. Alstroemeriaceae | 195 |
| 3.16. Pontederiaceae | 199 |
| 3.17. Trilliaceae | 202 |
| 3.18. Cupressaceae | 205 |
| 3.19. Podocarpaceae | 209 |
| 3.20. Grammitidaceae | 212 |
| 3.21. Marsileaceae | 216 |
| 3.22. Bryaceae | 219 |
| **4. Learning from the Holobaramin** | **223** |
| 4.1. Demography and Diversification | 223 |
| 4.2. Identifying Holobaramins | 228 |
| 4.3. Future Studies | 230 |
| **Appendix. A List of Identified Baramins** | **233** |
| **References** | **241** |
| **Index** | **251** |

# 1. Introduction

**1.1. Baraminology**

No creationist field over the past twenty years has advanced more rapidly than baraminology. Wise introduced baraminology at the 1990 International Conference on Creationism and two years later published a demonstration of its utility in a case study of the turtles (Wise 1990, 1992). In 1996, the BSG was formed (Frair 2000) and work began on the HybriDatabase (Wood et al. 2001). In 1997, the BSG published its first article, a biblical study of the Hebrew term *min* or "kind" (Williams 1997). In 1998, Robinson and Cavanaugh (1998a, 1998b) published two important studies that introduced statistical methods based on their "baraminic distance." In 1999, the first baraminology conference was held for a private audience of biologists, and in 2001 the first public conference was held. The first book on baraminology was published in 2003 (Wood and Murray 2003), and in 2005 a monograph on the Galápagos Islands significantly extended the number of baraminology case studies (Wood 2005a).

Despite this progress, much work remains to be done. The baraminology case studies are minuscule in number compared to the 1000 projected land animal baramins predicted by Jones (1973). These few case studies prevent important questions from being answered conclusively. For example, do holobaramins consist of a usual taxonomic rank? Do the baraminic distance methods give consistent results for holobaramins examined with different datasets? Do baraminic distance results reveal patterns consistent with discontinuity, or are such patterns the exception rather than the rule? To begin answering these questions, we must increase the number of baraminology case studies.

In this work, I apply statistical methods developed by Robinson and Cavanaugh (1998a) and myself (Wood 2005b, 2008) to 63 datasets taken from the published literature. Because this work is intended to increase the number of baraminology case studies rather than examine each dataset in taxonomic detail, I provide only a minimal interpretation of the results. I leave fuller, detailed descriptions of each group to experts in those fields. At the end of this work, I will try to draw some conclusions about baraminology and baraminic distance methods based on my findings.

## 1.2. Methods

Datasets were chosen according to the following two primary criteria: (1) The OTUs should be genera or species, preferably not higher ranks. This prevents making unwarranted assumptions about the continuity (or discontinuity) within OTUs. (2) The taxonomic scope of the dataset should cover an order, family, or subfamily. This allows testing of the common creationist assumption that the rank of family is approximately equivalent to the created kind. Accessibility also influenced my choice of datasets by biasing the sample towards recently-published datasets from prominent journals and datasets freely accessible from internet repositories.

The 63 datasets come from 39 animal (62.9%) and 22 plant (37.1%) groups. The animal datasets include 27 from vertebrates, eleven from arthropods, and one from annelids. Two families (Felidae and Fringillidae) are represented by multiple datasets. The plant datasets include seventeen from angiosperms, two each from gymnosperms and ferns, and one from the mosses.

BDISTMDS v. 2 was used to conduct a full baraminic distance correlation (BDC) analysis on each dataset. For technical details on the mathematics of the methods, consult other sources (e.g., Robinson and Cavanaugh 1998a; Wood 2005b). Datasets were first filtered according to character and taxic relevance. The cutoff values for each type of relevance varied according to the dataset and were chosen to balance high relevance with a large number of characters to be included in the baraminic distance calculations. Relevance cutoff values are included in the descriptions of the datasets.

Next baraminic distances, correlations, and probabilities were calculated for the original dataset and 100 bootstrap pseudoreplicates (Wood 2008). Finally, 3D multidimensional scaling (MDS) was calculated for the datasets. The ideal pattern of BDC showing a well-defined set of taxa united by significant, positive correlation and separated from the outgroup by significant, negative correlation appears in Figure 1. In practice, this pattern is rarely obtained. Alternative patterns include significant, negative correlation between taxa that are positively correlated with the same third taxon (e.g., horses; Cavanaugh et al. 2003); significant, positive correlation between ingroup and outgroup taxa (e.g., curculionids; Wood 2005a); and unexpected negative correlation between an otherwise easily definable taxonomic group (e.g. cormorants; Wood 2005a).

Interpretation of ambiguous BDC patterns is aided by 3D MDS, which allows visual examination of a representation of taxa in biological character space. In some cases, taxa of ambiguous affinity can be seen to be members of either the ingroup or outgroup based on their position in character space (Wood 2005b). In other cases, the MDS reveals a peculiar regularity, such as a tetrahedron or a line. Wood and Cavanaugh

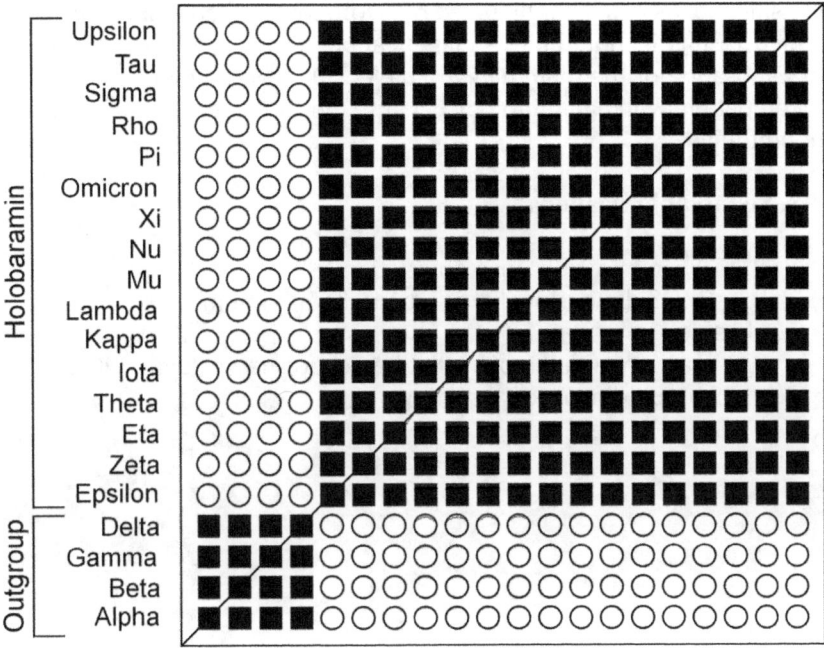

Figure 1. Hypothetical BDC results for an ideal holobaramin. Black squares denote significant, positive BDC, and open circles denote significant, negative BDC. The holobaramin and outgroups are indicated.

(2003) interpret lines with a chronological directionality as evidence of membership in a common baramin. The source of tetrahedral shapes is thought to be a bias (conscious or unconscious) towards characters that define groups in the dataset (Wood 2005a).

The results of the analyses are presented in a taxonomic order that blends modern with older taxonomic ordering. This same order is used in the appendix that lists published baraminological analyses. I have chosen to follow Linnaeus in listing animals first, with the primates first among them (Linnè 1806). The remaining ordering of mammalian orders follows roughly that of McKenna and Bell (1997). The arthropods are ordered with crustacea first, arachnids, and finally insects. The flowering plant classification follows Cronquist (1981). Remaining classifications were derived largely from the nonauthoritative taxonomy database of the National Center for Biotechnology Information (www.ncbi.nlm.nih.gov).

This ordering was deliberately chosen to emphasize the classically-recognized classification based on morphology and anatomy. Baraminology and the related concept of the cognitum emphasize the appearance and recognition of intuitive biological groups (Wood and Murray 2003; Sanders and Wise 2003). I recognize that modern classification is heavily influenced by evolutionary theory and cladistics, which results in nonintuitive classifications (such as placing the birds in Dinosauria or the mammals in Euteleosti). These would seem to be incompatible with baraminology because they are both evolutionary and nonintuitive.

**1.3. Key to Abbreviations**

BDC - baraminic distance correlation
$k_{min}$ - MDS dimensionality at which the minimum stress is observed
$F_{90}$ - Fraction of taxon pairs that have bootstrap values >90%.
MDS - multidimensional scaling

# 2. Animalia

## 2.1. Galagonidae (Vertebrata: Mammalia: Primates)

Dataset published by Masters & Brothers (2002)

| | |
|---|---|
| Characters in published dataset: | 36 |
| Taxa in published dataset: | 21 |
| Character relevance cutoff: | 0.95 |
| Characters used to calculate BD: | 36 |
| Taxa used in BDC and MDS analysis: | 21 |
| Stress for 3D MDS: | 0.103 |
| $k_{min}$: | 4 |
| Median bootstrap value: | 76 |
| $F_{90}$: | 0.219 |

Galagos are nocturnal primates from Africa also known as bushbabies. According to Masters and Brothers (2002), estimates of the number of galago species range from eleven in three genera to as many as forty different species. They list three phylogenetic questions that are unresolved: (1) the position of *Galagoides alleni*, (2) the relationship of the Zanzibar galagos (*Galagoides zanzibaricus* and *Galagoides granti*) with the rest of the family, and (3) the position of *Galago elegantulus*. To address these questions, they developed a morphological dataset of 36 craniodental characters from thirteen galagonid species and eight outgroup taxa. They classified their galagonid species into three genera: *Galago*, *Galagoides*, and *Otolemur*. In their analysis, they varied the outgroups and character weighting to produce seven different phylogenetic hypotheses. The position of all three questionable taxa listed above varied according to the outgroup chosen. With *Nyctocebus* or *Perodicticus* as outgroup, *Galagoides* species appear as monophyletic, but the *Galagoides* clade disappears with *Loris* or *Arctocebus* as outgroup. All of their trees conflicted to some extent with molecular trees obtained from mitochondrial DNA.

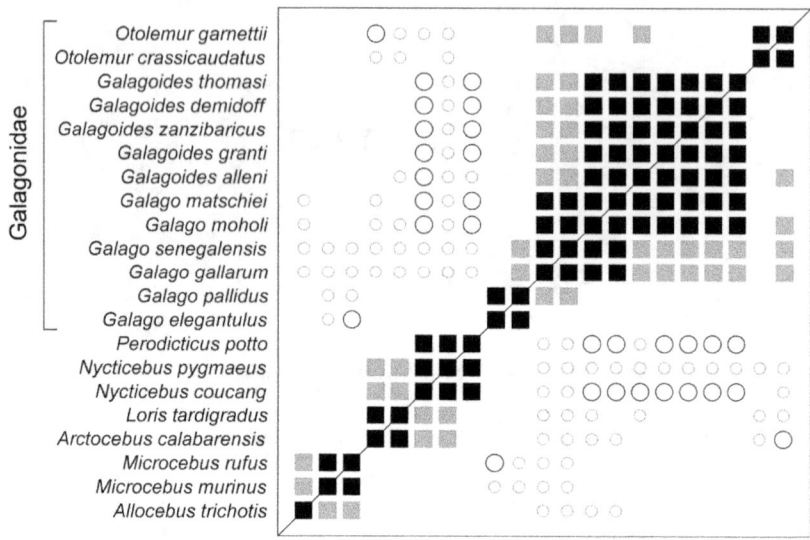

Figure 2. BDC bootstrap results for Galagonidae, as calculated by BDISTMDS (relevance cutoff 0.95). Closed square indicate significant, positive BDC; open circles indicate significant, negative BDC. Black symbols indicate bootstrap values >90% in a sample of 100 pseudoreplicates. Grey symbols represent bootstrap values ≤90%.

For the baraminological study, all taxa and characters were retained after relevance filtering (see above). BDC results show signficant, positive BDC between most *Galago* and *Galagoides* species (Figure 2). Two galagonid groups appear as outliers in these results: *Otolemur* and the species *Galago pallidus* and *Galago elegantulus*. *Otolemur crassicaudatus* correlates positively with only *O. gamettii*, but *O. gamettii* correlates positively with species of *Galago* and *Galagoides*. *Galago elegantulus* correlates positively with only *Galago pallidus*, but *pallidus* in turn correlates positively two other *Galago* species (*gallarum* and *senegalensis*). Thus, all the galagonids are linked together by significant, positive BDC. Compared to the outgroup loris species, galagonids have only significant, negative BDC. Of the 104 ingroup-outgroup taxon pairs, 54 have significant, negative BDC and the rest have nonsignificant correlation.

The bootstrap values are quite high for this dataset. Of the 44 galagonid pairs that share significant, positive BDC, 28 have bootstrap values >90%. Bootstrap values are lower for the loris-galago taxon pairs, with only 15 of the 54 taxon pairs with significant, negative BDC having bootstrap values >90%.

Multidimensional scaling revealed two layers of taxa in three dimensions (Figure 3). The two layers correspond to the galagos and the outgroup taxa. The low 3D stress and dimensions of minimal stress (see above) indicate that the baraminic distances are well-represented in

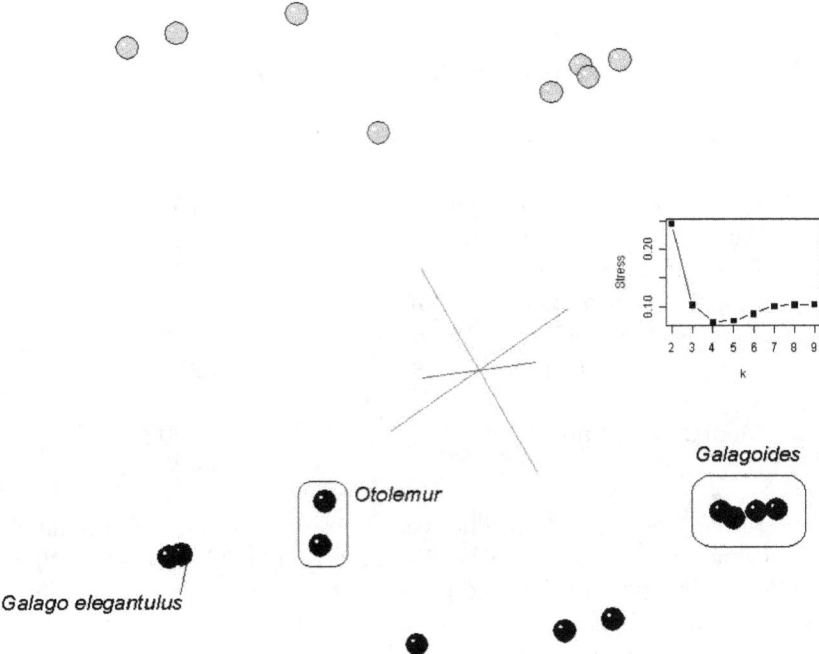

Figure 3. Three-dimensional MDS applied to Galagonidae baraminic distances and the stress of k-dimensional MDS on the same baraminic distance matrix plotted as a function of the number of dimensions (k). Galagonids are shown in black and outgroup taxa in gray.

the 3D MDS results. Within the Galagonidae, the *Galagoides* species, including *alleni*, cluster together, as do the *Otolemur* species. The *Galago* species are more diffuse with *elegantulus* and *pallidus* outlying from the center of the galagonid cluster. The outgroup taxa are separated from the cluster of galagonid taxa.

Based on the significant, positive BDC within the Galagonidae and the signficant, negative BDC between galagonids and outgroup taxa, I would classify the galagonids in a single holobaramin. This classification should be considered provisional, since the data here examined is only craniodental and therefore not holistic. The BDC and MDS seem to resolve the position of the *Galagoides* species fairly well, including the questionable Zanzibar species and *Galagoides alleni*. The grouping of *Galago elegantulus* and *Galago pallidus* is also consistent with Masters and Brothers (2002), who found that clade in all phylogenetic analyses. The ambiguity of placement within the family's phylogeny seems to be in part caused by its distance from the rest of the galagonid species and the diffuseness of the genus *Galago*.

## 2.2. Dasypodidae (Vertebrata: Mammalia: Xenarthra)

Dataset published by Gaudin & Wible (2006)

| | |
|---|---|
| Characters in published dataset: | 163 |
| Taxa in published dataset: | 21 |
| Character relevance cutoff: | 0.9 |
| Characters used to calculate BD: | 74 |
| Taxa used in BDC and MDS analysis: | 20 |
| Stress for 3D MDS: | 0.271 |
| $k_{min}$: | 11 |
| Median bootstrap value: | 90 |
| $F_{90}$: | 0.474 |

Armadillos are a distinctively armored group of mammals consisting of 21 extant species in eight genera (Gaudin and Wible 2006). Extinct glyptodonts and pampatheres have also been associated with the armadillos in the suborder Cingulata. Gaudin and Wible (2006) include the pampatheres in the armadillo family Dasypodidae and place the glyptodonts in a separate family Glyptodontidae. Their morphological dataset consists of 163 craniodental characters scored for 21 taxa. The taxa include two outgroups, the anteater (*Tamandua*) and sloth (*Bradypus*), as well as sixteen dasypodids. The remaining taxa are the horned armadillo *Peltephilus*, the pampathere *Vassallia*, and the glyptodont *Propalaeohoplophorus*. Due to an error in the printed dataset, *Cabassous* was omitted from the baraminic distance calculations.

Baraminic distances were calculated at a relevance cutoff of 0.9, which allows for only two unknown character states for this few taxa. Of the 163 characters input into BDISTMDS, only 74 were used to calculate baraminic distances. BDC results reveal a block of fourteen taxa representing the bulk of the dasypodids and the pampathere *Vassallia* (Figure 4). Generally, taxon pairs within this block share significant, positive BDC. Peripheral to this block are two taxa that are significantly, positively correlated with only three other taxa, the glyptodont *Propalaeohoplophorus* and the nine-banded armadillo *Dasypus*. Three other ingroup taxa show no significant, positive BDC with any other taxa: *Stegotherium*, the horned armadillo *Peltephilus*, and the giant armadillo *Priodontes*. *Stegotherium* actually shows significant, negative BDC with nine other cingulate taxa but not with the outgroup taxa. The outgroup taxa have mostly significant, negative BDC with members of the large block of dasypodids and *Vassallia*. Bootstrap support is moderately

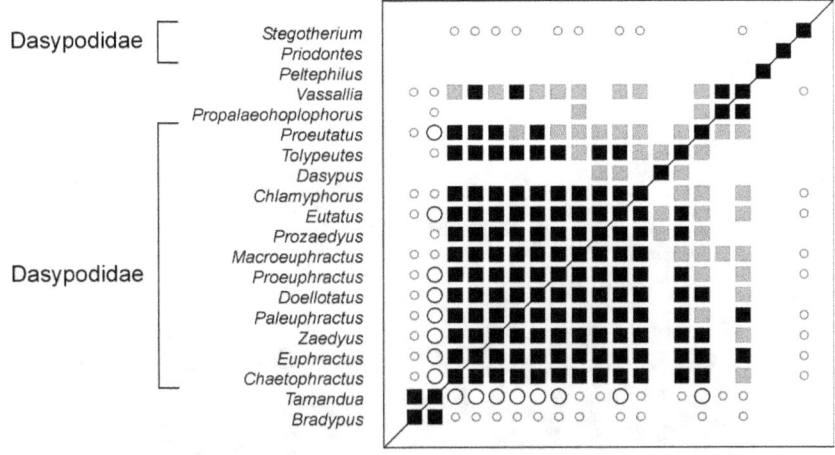

Figure 4. BDC bootstrap results for Dasypodidae, as calculated by BDISTMDS (relevance cutoff 0.9). Closed square indicate significant, positive BDC; open circles indicate significant, negative BDC. Black symbols indicate bootstrap values >90% in a sample of 100 pseudoreplicates. Grey symbols represent bootstrap values ≤90%.

high, but especially high (>90%) within a core group of ten dasypodids consisting of *Chlamyphorus, Eutatus, Prozaedyus, Macroeuphractus, Proeuphractus, Doellotatus, Paleuphractus, Zaedyus, Euphractus,* and *Chaetophractus.*

The MDS reveals a complicated pattern of taxa (Figure 5), made more complicated by the unusually high 3D stress, 0.271. *Priodontes*, which shares no significant, positive BDC with other cingulates, nevertheless appears closely allied with *Dasypus*, which is positively correlated with three other cingulates. *Peltephilus* and *Stegotherium* are more distant from the other taxa and from each other, which confirms their lack of significant, positive BDC with other cingulates. The two outgroup taxa are most distant of all from the main cloud of cingulate taxa.

Based on the significant, positive BDC that unites the bulk of the dasypodids and *Vassallia*, that group is likely to be a monobaramin. More problematic is the position of the giant armadillo *Priodontes*, which has only nonsignificant BDC when compared to any other taxon in the dataset. *Priodontes* is unmistakably an armadillo, but these characters make it at best an outlying armadillo. The significant, negative BDC between the dasypodid + *Vassallia* monobaramin and members of the outgroup suggest that this group of armadillos is also a holobaramin. Due to lack of significant BDC, the glyptodont *Propalaeohoplophorus* cannot be excluded from or included in the dasypodid + *Vassallia* holobaramin. The significant, negative BDC between *Stegotherium* and

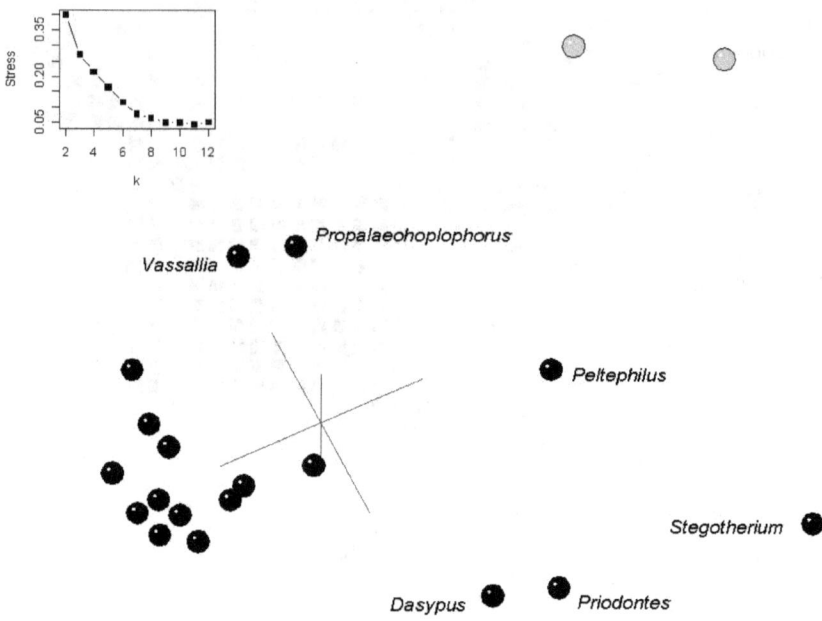

Figure 5. Three-dimensional MDS applied to Dasypodidae baraminic distances and the stress of k-dimensional MDS on the same baraminic distance matrix plotted as a function of the number of dimensions (k). Dasypodids and other cingulates are shown in black. Outgroup taxa are shown in gray.

taxa in the dasypodid + Vassallia holobaramin suggest a discontinuity placing *Stegotherium* in a different holobaramin.

There is much room for uncertainty in these conclusions. First, the entirely craniodental dataset is not holistic, and thus is not optimal for baraminology. Second, the high 3D stress and the eleven dimensions needed to minimize the stress suggest that the baraminic distances are quite complex and that the 3D MDS do not reliably represent the full complexity of the baraminic distances. It is possible that the pampathere *Vassallia* is in a separate holobaramin from the armadillos, and it is very likely that the giant armadillo *Priodontes* should be included in the same holobaramin as the remaining armadillos. Thus, the most conservative conclusion I could draw from these results is that the dasypodid + *Vassallia* group is a monobaramin and likely part of a larger holobaramin. There does seem to be discontinuity between the outgroups and the cingulates and between *Stegotherium* and other cingulates, as evidenced by the significant, negative BDC.

## 2.3. Leporidae (Vertebrata: Mammalia: Lagomorpha)

Dataset published by Wible (2007)

| | |
|---|---|
| Characters in published dataset: | 59 |
| Taxa in published dataset: | 13 |
| Character relevance cutoff: | 0.95 |
| Characters used to calculate BD: | 40 |
| Taxa used in BDC and MDS analysis: | 11 |
| Stress for 3D MDS: | 0.037 |
| $k_{min}$: | 3 |
| Median bootstrap value: | 95 |
| $F_{90}$: | 0.655 |

The 61 species of rabbits and hares are classified into eleven extant genera. Though they look like rodents, they are actually classified in a separate order Lagomorpha. The other lagomorph family, Ochotonidae, consists of 30 extant species of pikas in a single genus. In his study of rabbit and pika cranial osteology, Wible (2007) created a dataset of 59 craniomandibular characters and 13 taxa. The taxa included only six leporid genera, a single extant pika *Ochotona princeps*, one fossil pika *Prolagus sardus*, and the "stem lagomorphs" *Gomphos elkema* and *Palaeolagus haydeni*. The nonlagomorph outgroup taxa are *Rhombomylus*, *Mimolagus*, and *Mimotona*, rodent-like fossils closely allied to the lagomorphs. According to Wible's phylogeny, both Ochotonidae and Leporidae are monophyletic, and the Lagomorpha excludes *Gomphos* and *Palaeolagus*.

Taxa with taxic relevance less than 0.78 (*Mimolagus* and *Mimotona*) were removed from the dataset to improve the number of characters used for baraminic distance calculations. At a character relevance of 0.95, only 40 of the 59 characters were used to calculated baraminic distances. The BDC results show significant, positive BDC between all possible leporid taxon pairs (Figure 6). The two ochotonid taxa are also positively correlated with each other but not significantly correlated with any other taxa in the dataset. Two of the outgroup taxa, *Gomphos* and *Rhombomylus*, are negatively correlated with the leporids. Bootstrap results are very high for all significant correlations (median 95%).

The MDS results were extraordinarily good, with a minimal stress of only 0.037 at three dimensions. The clustering pattern unsurprisingly show a tight cluster of leporids separated from the outgroup and ochotonid taxa (Figure 7). Most distant from the leporids are *Gomphos*

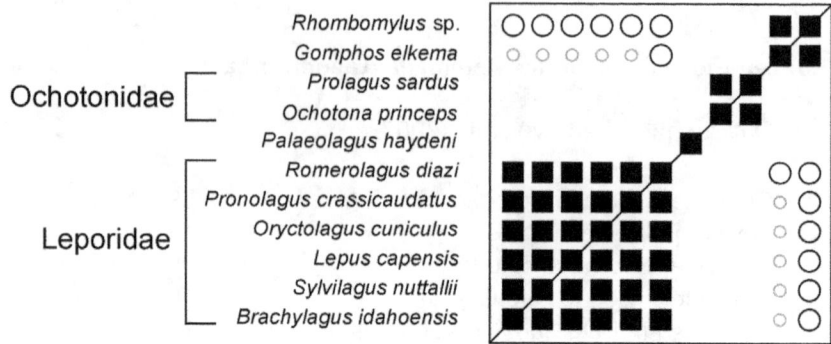

Figure 6. BDC bootstrap results for the Leporidae dataset, as calculated by BDISTMDS (relevance cutoff 0.95). Closed square indicate significant, positive BDC; open circles indicate significant, negative BDC. Black symbols indicate bootstrap values >90% in a sample of 100 pseudoreplicates. Grey symbols represent bootstrap values ≤90%.

and *Rhombomylus*, with *Palaeolagus* in an intermediate position between the leporids and the other outgroups.

Based on these results, the Leporidae appear to be a monobaramin based on the significant, positive BDC and the close MDS clustering. It might be tempting to declare that the ochotonids are also a monobaramin, but since the pikas are poorly represented in this dataset, such a conclusion would be premature. The significant, negative BDC between the leporids and two of the outgroup taxa suggests that the leporids are a holobaramin. Since the dataset is only craniomandibular and represents only six of the eleven extant lagomorph genera, the leporid holobaramin should be considered preliminary until it can be confirmed with a more holistic dataset that includes a wider sampling of leporid taxa.

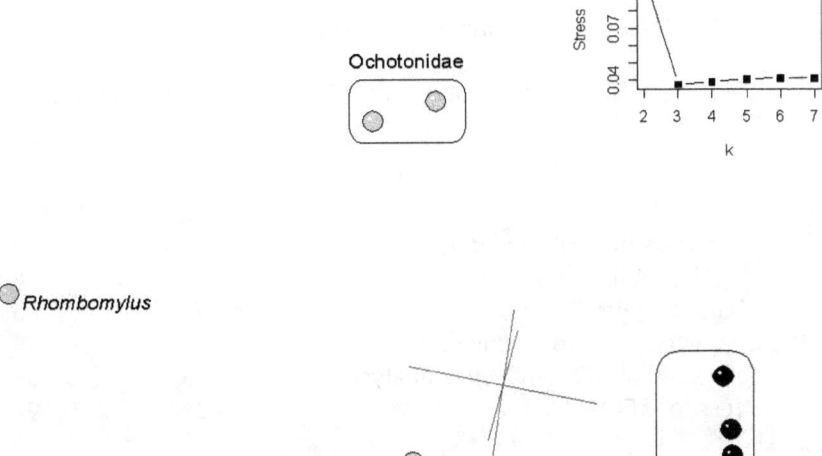

Figure 7. Three-dimensional MDS applied to Leporidae baraminic distances and the stress of $k$-dimensional MDS on the same baraminic distance matrix plotted as a function of the number of dimensions ($k$). Leporids are shown in black and outgroup taxa in gray.

## 2.4. Felidae (Vertebrata: Mammalia: Carnivora)

Datasets published by Mattern and McLennan (2000) and Salles (1992)

|  | Mattern & McLennan | Salles |
|---|---|---|
| Characters in published dataset: | 68 | 57 |
| Taxa in published dataset: | 37 | 38 |
| Character relevance cutoff: | 0.9 | 0.95 |
| Characters used to calculate BD: | 55 | 48 |
| Taxa used in BDC and MDS analysis: | 37 | 38 |
| Stress for 3D MDS: | 0.214 | 0.194 |
| $k_{min}$: | 8 | 8 |
| Median bootstrap value: | 72 | 68 |
| $F_{90}$: | 0.273 | 0.275 |

The cats were analyzed previously in one of the earliest BDC studies by Robinson and Cavanaugh (1998b). They concluded that Felidae was a holobaramin based on both hybridization and BDC. Using a holistic dataset that they compiled, they found significant, positive BDC within the Felidae and significant, negative BDC for felid-outgroup taxon pairs. They used the hyena (*Crocuta crocuta*) and meerkat (*Suricata suricatta*) as outgroups. This was the first holobaramin identified using BDC.

I used two different datasets to reanalyze the felids. Salles's (1992) study consisted of 57 characters, 44 of which are craniodental. Only 38 felid taxa were studied, no outgroup taxa were included in the dataset. The more recent analysis of Mattern and McLennan (2000) was based on Salles's dataset but included an additional ten karyological characters and one morphological character. They also used the hyena and mongoose (*Galidia elegans*) as outgroup taxa. Since baraminology emphasizes holistic data, neither of these datasets that focus primarily on craniodental data are ideal for baraminological study.

The BDC results for the Salles dataset reveal two well-defined groups (Figure 8). The genera *Uncia*, *Panthera*, and *Neofelis* (six species in all) comprise one group, and all other species fall into the other. Six taxon pairs with significant, positive BDC connect the two groups (*Uncia/Leopardus wiedii*, *Uncia/Leopardus pardalis*, *Neofelis/Leopardus wiedii*, *Neofelis/Leopardus pardalis*, *Neofelis/Profelis aurata*, *Neofelis/Puma concolor*). Significant, negative BDC was observed between the *Uncia/Panthera/Neofelis* group and *Otocolobus manul* and the genera *Felis* and *Lynx*. The cheetah (*Acinonyx jubatus*) had no significant, positive BDC with other felids but significant, negative BDC with four *Felis* species (*margarita, nigripes, sylvestris, bieti*), *Leopardus jacobita, Prionailurus rubiginosa,* and *Otocolobus manul*. Bootstrapping results were moderate, with highest bootstrap values (>90%) observed within genera *Felis, Panthera, Lynx,* and *Leopardus*. None of the signficant, negative BDC had bootstrap values >90%. The MDS results for the Salles dataset generally confirmed the findings of the BDC analysis (Figure 9). As expected, the cheetah (*Acinonyx*) and the *Uncia/Panthera/Neofelis* group were notable outliers from the main, compact cluster of felid taxa.

Expansion of the dataset by Mattern and McLennan resulted in very similar BDC results (Figure 10). The same two groups are visible, as is the outlying cheetah. The outgroups are positively correlated with the *Uncia/Panthera/Neofelis* group as well as species of the genera *Profelis, Leopardus, Ictailurus,* and *Puma*. The outgroups are not negatively correlated with any taxa. The median bootstrap value increased slightly, but the fraction of taxon pairs with >90% bootstrap values remained essentially unchanged (0.275 for Salles vs. 0.273 for Mattern and McLennan). Five taxon pairs with significant, negative BDC had bootstrap values >90%: *Otocolobus/Panthera leo, Otocolobus/Panthera onca, Otocolobus/Panthera pardus, Otocolobus/Panthera tigris,* and *Panthera uncia/Felis nigripes*. Not surprisingly, the 3D MDS pattern closely resembles that of the Salles dataset (Figure 11). The cheetah and *Uncia/Panthera/Neofelis* group remain as outliers from the main cluster of cat taxa. The hyena and mongoose outgroup taxa appear in an intermediate position between the main cluster of cats and the *Uncia/Panthera/Neofelis* group.

Based on the results of the Salles dataset alone, I would conclude that there is no discontinuity within the cats. The groups outlying from the main cluster of cats have significant, negative BDC with only a few of the other cats, and the negative correlation between the *Uncia/Panthera/Neofelis* group and the genera *Felis* and *Lynx* are offset by significant positive BDC between *Uncia/Panthera/Neofelis* and other cat taxa. This conclusion should be considered tentative, though, since the data is entirely craniodental and not holistic. The additional characters and outgroup taxa in the Mattern and McLennan version of the same dataset do not alter these basic conclusions. The larger dataset does suggest that there is no discontinuity between felids and nonfelids, but again since the dataset is not holistic, evident continuity with outgroup taxa may be artifactual. Because Robinson and Cavanaugh (1998b) included the Salles dataset in their more holistic 287-character dataset, I would not reject their conclusion that the cats represent one holobaramin, discontinuous from nonfelid carnivores.

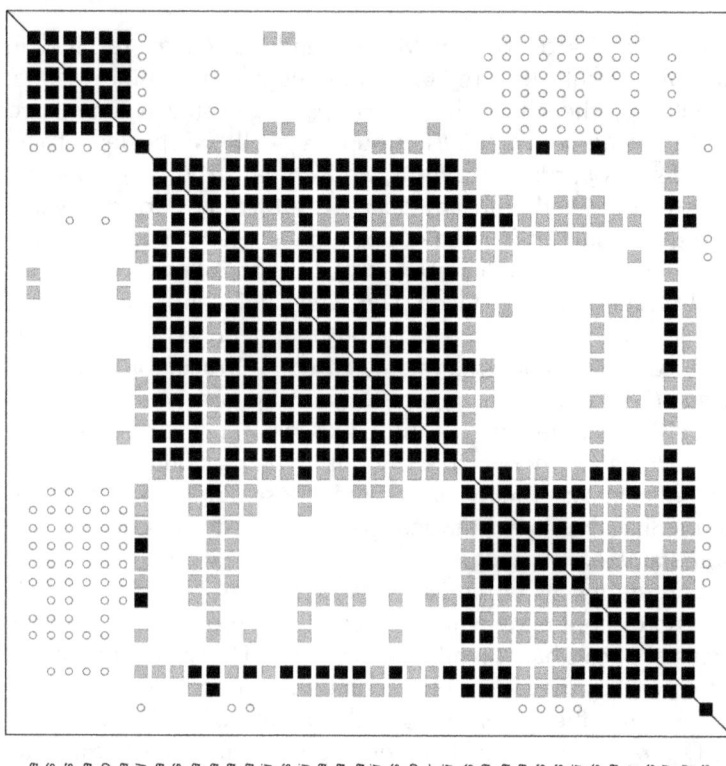

Figure 8. BDC bootstrap results for the Salles Felidae dataset, as calculated by BDISTMDS (relevance cutoff 0.95). Closed square indicate significant, positive BDC; open circles indicate significant, negative BDC. Black symbols indicate bootstrap values > 90% in a sample of 100 pseudoreplicates. Grey symbols represent bootstrap values ≤ 90%.

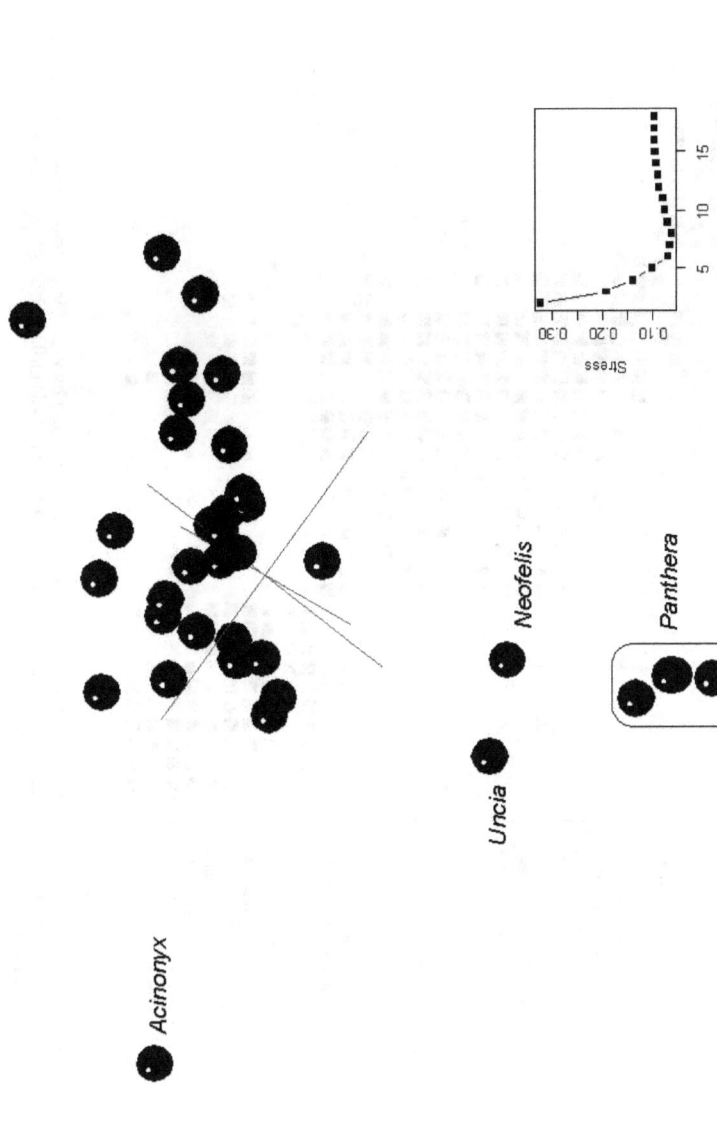

Figure 9. Three-dimensional MDS applied to Felidae baraminic distances from the Salles dataset. Also shown is the stress of $k$-dimensional MDS on the same baraminic distance matrix plotted as a function of the number of dimensions ($k$).

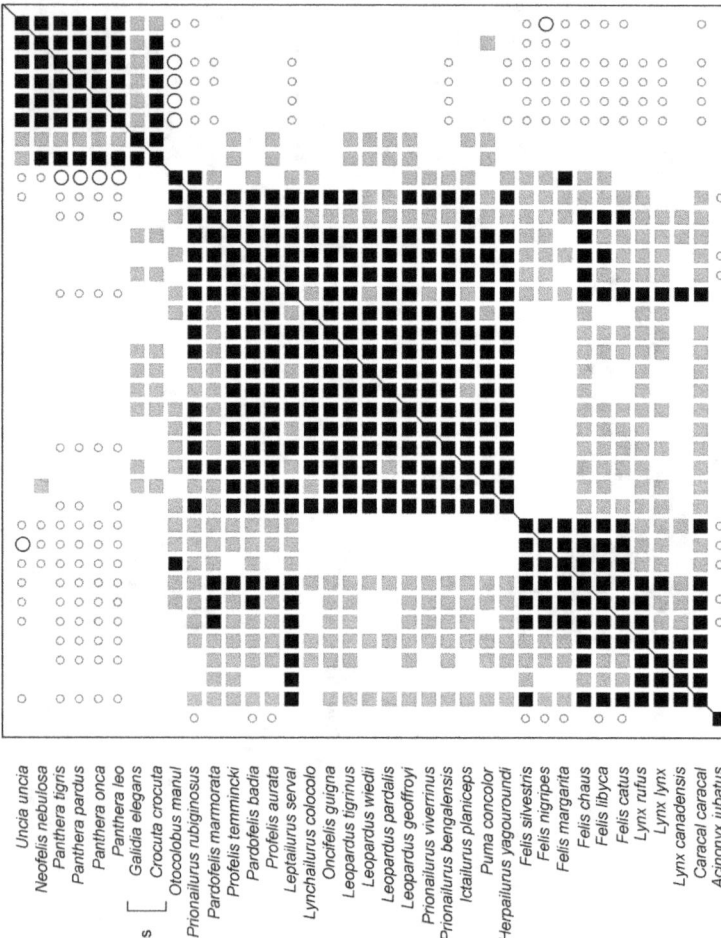

Figure 10. BDC bootstrap results for the Mattern and McLennan Felidae dataset, as calculated by BDISTMDS (relevance cutoff 0.9). Closed square indicate significant, positive BDC; open circles indicate significant, negative BDC. Black symbols indicate bootstrap values > 90% in a sample of 100 pseudoreplicates. Grey symbols represent bootstrap values ≤90%.

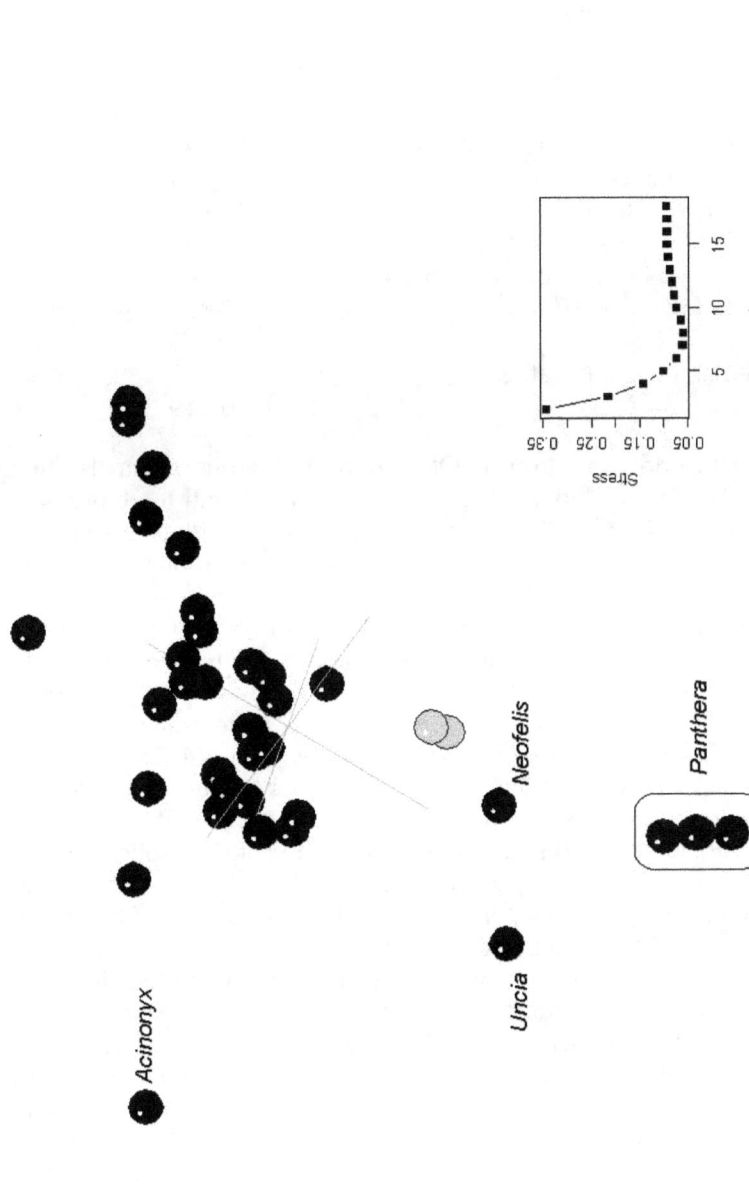

Figure 11. Three-dimensional MDS applied to Felidae baraminic distances from the Mattern and McLennan dataset. Also shown is the stress of $k$-dimensional MDS on the same baraminic distance matrix plotted as a function of the number of dimensions ($k$). Outgroup taxa are shown in grey.

## 2.5. Viverridae (Vertebrata: Mammalia: Carnivora)

Dataset published by Gaubert et al. (2005)

| | |
|---|---|
| Characters in published dataset: | 349 |
| Taxa in published dataset: | 44 |
| Character relevance cutoff: | 0.95 |
| Characters used to calculate BD: | 252 |
| Taxa used in BDC and MDS analysis: | 44 |
| Stress for 3D MDS: | 0.338 |
| $k_{min}$: | 15 |
| Median bootstrap value: | 92 |
| $F_{90}$: | 0.519 |

The viverrids are a group of Old World carnivorous mammals allied with the Felidae, including civets, genets, linsangs, and the binturong. Previously classified in Viverridae, the Malagasy viverrid-like carnivores (inlcuding the fossa, *Cryptoprocta ferox*) and the mongooses have recently been removed from Viverridae based on molecular evidence (see Gaubert et al. 2005). The viverrid-like carnivores and mongooses are now classified in family Eupleridae. The phylogeny of the viverridae and viverrid-like carnivores was examined by Gaubert et al. (2005) using a dataset of 349 characters, of which 221 are craniodental, 57 postcranial, and 71 morphology/soft anatomy. The 44 taxa included 28 of the approximately 35 viverrid species, four euplerids, and twelve carnivore outgroup species.

For BDC, the 0.95 character relevance cutoff resulted in the omission of 97 characters, leaving 209 craniodental, 25 postcranial, and 18 morphology/soft anatomy characters for calculation of baraminic distances. The BDC results show a single block of 33 taxa characterized by extensive significant, positive BDC (Figure 12). This block includes all the viverrids *sensu stricto*, the euplerids *Fossa fossana* and *Eupleres goudotii*, and the African palm civet *Nandinia binotata*. The fossa *Cryptoprocta* shows significant, positive BDC with *Fossa*, *Eupleres*, and *Nandinia*, as well as the mongoose *Galidia*, which is in turn positively correlated with two herpestids *Herpestes* and *Mungos*. The remaining outgroup taxa show only significant, negative correlation with members of the main block of viverrid taxa. Bootstrap values are fairly high for these results (see table), especially within the main block of 33 viverrid and viverrid-like taxa.

The MDS results reveal two clusters of viverrids closely allied with the viverrid-like euplerids (Figure 13). The taxa appear as a diffuse cloud of points in character space. The herpestids cluster closely with *Galidia*, and *Nandinia* occupies a position very near the euplerids and viverrids. Consistent with the apparent lack of significant, negative BDC, there appear to be no substantial gaps or well-defined groups of taxa. The non-euplerid, non-viverrid taxa appear as a arc, beginning with *Nandinia* close to the viverrids and terminating with the felids *Felis* and *Lynx*.

These results are difficult to interpret baraminologically. First, the dataset is far from holistic, with 209 out of 252 (83%) of the characters being craniodental. Based on the BDC results alone, Viverridae plus *Prionodon, Eulperes, Fossa*, and *Nandinia* (hereafter Viverridae *sensu lato*) could be classified as a monobaramin based on extensive significant, positive BDC. Inclusion of *Cryptoprocta, Galidia*, and the two herpestids in this monobaramin would be possible but much more tentative. Lack of correlation with other carnivore outgroup taxa should not be interpreted as evidence of discontinuity, and therefore no holobaramin can be estimated from these results. The MDS results confirm these conclusions. *Galidia* and the two herpestids are not in the cluster of viverrid taxa, and thus suggest that they are not part of Viverridae *sensu lato*. The lack of any obvious gaps in the MDS clustering pattern is consistent with the lack of significant, negative BDC; all of which is consistent with a lack of evidence for discontinuity.

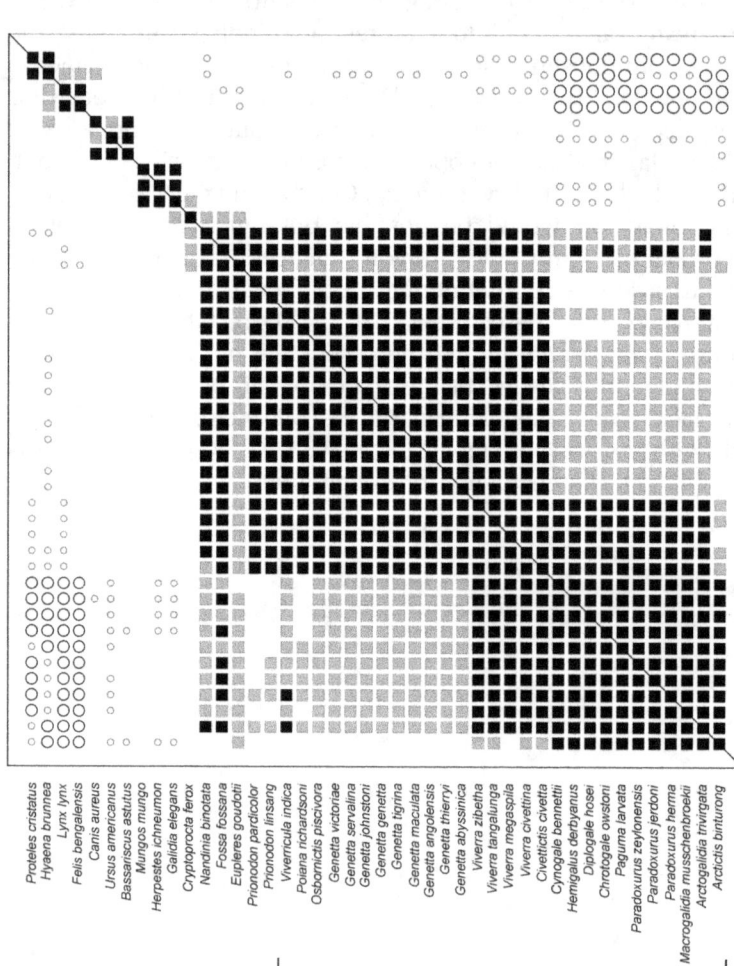

Figure 12. BDC bootstrap results for Viverridae, as calculated by BDISTMDS (relevance cutoff 0.95). Closed square indicate significant, positive BDC; open circles indicate significant, negative BDC. Black symbols indicate bootstrap values > 90% in a sample of 100 pseudoreplicates. Grey symbols represent bootstrap values ≤ 90%.

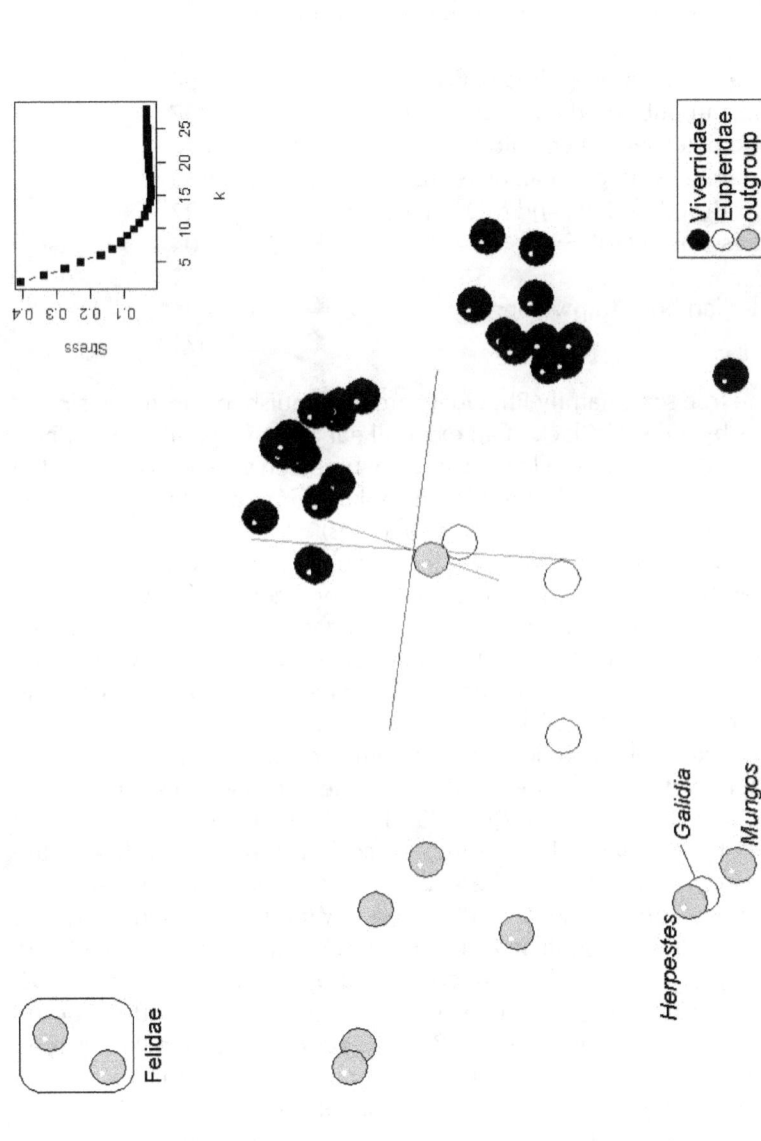

Figure 13. Three-dimensional MDS applied to Viverridae baraminic distances and the stress of *k*-dimensional MDS on the same baraminic distance matrix plotted as a function of the number of dimensions (*k*). Viverrids are shown in black, euplerids in white, and outgroup taxa in gray.

### 2.6. Phocidae (Vertebrata: Mammalia: Carnivora)

Dataset published by Bininda-Emonds & Russell (1996)

| | |
|---|---|
| Characters in published dataset: | 196 |
| Taxa in published dataset: | 27 |
| Character relevance cutoff: | 0.95 |
| Characters used to calculate BD: | 189 |
| Taxa used in BDC and MDS analysis: | 27 |
| Stress for 3D MDS: | 0.204 |
| $k_{min}$: | 12 |
| Median bootstrap value: | 96 |
| $F_{90}$: | 0.621 |

The true seals (family Phocidae) are distinguishable from the similar sea lions by the seals' lack of an external ear and the inability to support their weight on their hind legs. There are nineteen species of seals in ten genera. Seals are generally found in the polar regions of both hemispheres, except for the monk seals (genus *Monachus*) which are (or were) found in the Mediterranean, Hawaii, and the Caribbean. The nearest outgroups are generally thought to be the walrus (Odobenidae: *Odobenus rosmarus*) and the sea lions (Otariidae). The phylogenetic analysis of Bininda-Emonds and Russell (1996) included all seal species plus eight outgroup taxa. They sampled 196 characters (129 craniomandibular, 23 dental, 41 postcranial skeletal, and 3 miscellaneous).

All taxa were retained for baraminic distance calculations, and a character relevance cutoff of 0.95 resulted in the omission of only seven characters (1, 2, 86, 90, 123, 192, 193: five craniomandibular and two postcranial). Two groups of seals are readily visible in the BDC results (Figure 14). These groups correspond to the two seal subfamilies Monachinae and Phocinae. Within each subfamily, all taxon pairs have significant, positive BDC with >90% bootstrap values. The two subfamilies are connected by significant, positive BDC between the phocine hooded seal (*Cystophora cristata*) and six of the monachines. Significant, positive BDC also occurs between the bearded seal (Phocinae: *Erignathus barbatus*) and three of the monachines. No phocine/monachine taxon pairs share significant, negative BDC. All outgroup taxa have significant, negative BDC with all members of at least one subfamily, and two taxa, the bear (*Ursus*) and wolf (*Canis*), correlate negatively with all phocids. No outgroup taxon has significant, positive

BDC with any phocid. Overall, the bootstrap values are extremely good (median 96%).

The MDS results are not as good as the BDC, with a 3D stress of 0.2 and a minimum stress of 0.03 at twelve dimensions. Nevertheless, the 3D MDS results correspond well to the groups observed in the BDC analysis (Figure 15). The Phocinae and Monachinae form two clusters with *Cystophora* in between. The outgroup taxa are in a separate cluster, with the walrus (*Odobenus*) and the California sea lion (*Zalophus*) nearest the phocids.

Given the holistic nature of the character sample, the large number of characters, and the strong BDC results, the classification of Phocidae as a holobaramin seems indisputable. The phocids are united by continuity, as evidenced by extensive positive BDC, and are separated from other taxa by discontinuity, as evidenced by extensive negative BDC with outgroup taxa. There is no spurious negative BDC within the Phocidae nor any positive BDC between a phocid and an outgroup taxon. Phocidae is a holobaramin.

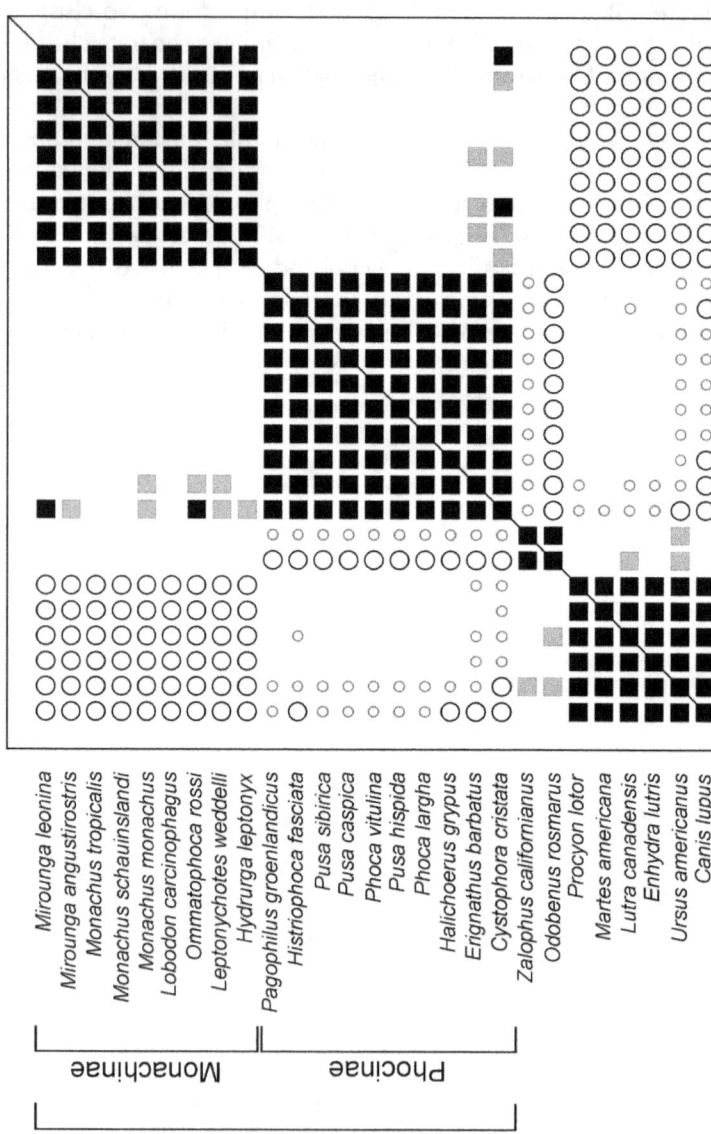

Figure 14. BDC bootstrap results for Phocidae, as calculated by BDISTMDS (relevance cutoff 0.95). Closed square indicate significant, positive BDC; open circles indicate significant, negative BDC. Black symbols indicate bootstrap values >90% in a sample of 100 pseudoreplicates. Grey symbols represent bootstrap values ≤90%.

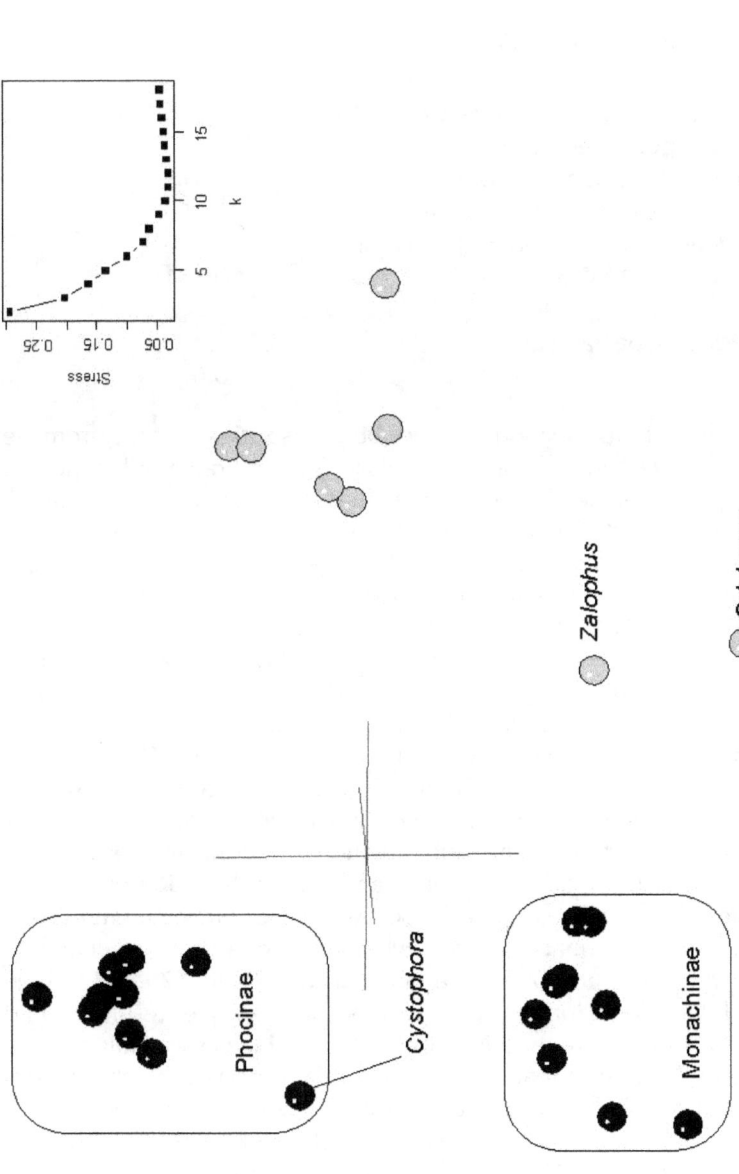

Figure 15. Three-dimensional MDS applied to Phocidae baraminic distances and the stress of *k*-dimensional MDS on the same baraminic distance matrix plotted as a function of the number of dimensions (*k*). Phocids are shown in black and outgroup taxa in gray.

## 2.7. Erinaceidae (Vertebrata: Mammalia: Erinaceomorpha)

Dataset published by Gould (1997)

| | |
|---|---|
| Characters in published dataset: | 135 |
| Taxa in published dataset: | 47 |
| Character relevance cutoff: | 0.9 |
| Characters used to calculate BD: | 68 |
| Taxa used in BDC and MDS analysis: | 25 |
| Stress for 3D MDS: | 0.151 |
| $k_{min}$: | 5 |
| Median bootstrap value: | 98 |
| $F_{90}$: | 0.677 |

The hedgehogs and moonrats are about 24 species primarily from the Old World, although fossil species are also known from North America. Subfamily Erinaceinae, the spiny hedgehogs, are found in Europe, Africa, and the Middle East. Subfamily Hylomyinae, the moonrats, are restricted to southeast Asia. At least one fossil hedgehog is known from North America, and fossil moonrats have been found in Africa, Europe, and North America. At least one fossil form, the North American *Proterix* has been classified variously in the Erinaceinae or Hylomyinae. Two other erinaceid subfamilies, Tupaiodontinae and Brachyericinae, are known from the fossil record of Asia and North America respectively.

Gould's morphological dataset (appendix 22 in Gould 1997) contains 135 characters scored for 47 taxa. The characters include 61 cranial, 59 dental characters, nine nonskeletal characters, and six postcranial characters. The taxa include 45 erinaceids and three outgroup taxa (*Eolestes*, tenrecoids and soricoids). Representation of erinaceid diversity is quite good, with two brachyericines, one tupaiodontine, 22 erinaceines, 17 hylomyines, one taxon of uncertain affinity (*Litolestes*), and one undescribed taxon (RGM 179.327, which clustered with the hylomyines in Gould's analysis). Her results (Gould 1995, 1997) support monophyly for the Erinaceidae as a whole and for the subfamilies Erinaceinae and Hylomyinae. The two brachyericines were also monophyletic and likely the sister taxon to the Erinaceinae.

Taxa had to be removed from the dataset before baraminic distances could be calculated. All taxa with taxic relevance less than 0.5 were eliminated from the dataset. These eliminations left a dataset of 25 taxa with fourteen erinaceines, eight hylomyines, one brachyericine (*Brachyerix*) and two outgroups (tenrecoids and soricoids). The resulting dataset still retained only 68 out of 135 characters after filtering at a relevance cutoff of 0.9. The remaining dataset consisted of 46 cranial and 22 dental characters. All postcranial and nonskeletal characters were omitted.

BDC results reveal two blocks of taxa and two outliers (Figure 16). The first block of taxa contains all the Erinaceinae except *Amphechinus edwardsi*, and the second block contains the hylomyines and the outgroups. The outliers are *Amphechinus edwardsi* and the brachyericine *Brachyerix macrotis*. In Gould's (1997) phylogeny, *Amphechinus edwardsi* is the most basal erinaceine, but its congener *Amphechinus rusingensis* (here omitted due to missing data) was not its sister taxon, thus indicating that the genus *Amphechinus* is polyphyletic. Within the Erinaceinae (- *Amphechinus*), all taxon pairs share significant, positive BDC with bootstrap values >90%. Bootstrap values for the outgroup + hylomyine group are not as good, and eight taxon pairs out of 45 do not have any significant BDC. Every taxon of the Erinaceinae (- *Amphechinus*) shows significant, negative BDC when compared to any taxon of the outgroup + hylomyine group. *Brachyerix* is not significantly correlated with any other taxa, and *Amphechinus* is correlated negatively only with a single hylomyine, *Hylomys suillus*.

Not surprisingly, the MDS results show *Amphechinus* separated from a tight cluster of other Erinaceinae taxa (Figure 17). The hylomyines are more dispersed than the erinaceines, with both *Amphechinus* and *Brachyerix* positioned between them. The outgroup tenrecoids and soricoids are separated from the hylomyines by a gap but are closer to the hylomyines than to the erinaceines.

Based on the tight clustering pattern in MDS and the significant, positive BDC, Erinaceinae is definitely a monobaramin. The negative BDC correlation with the hylomyines also supports a discontinuity around the Erinaceinae, making it a holobaramin. The ambiguous positioning of the *Brachyerix* and *Amphechinus* might suggest a connection between the hylomyines and erinaceines, but at this stage such a conclusion is unwarranted. Erinaceinae seems to be a well-supported holobaramin. Only the nonholistic nature of the character set renders these conclusions less certain than they could be.

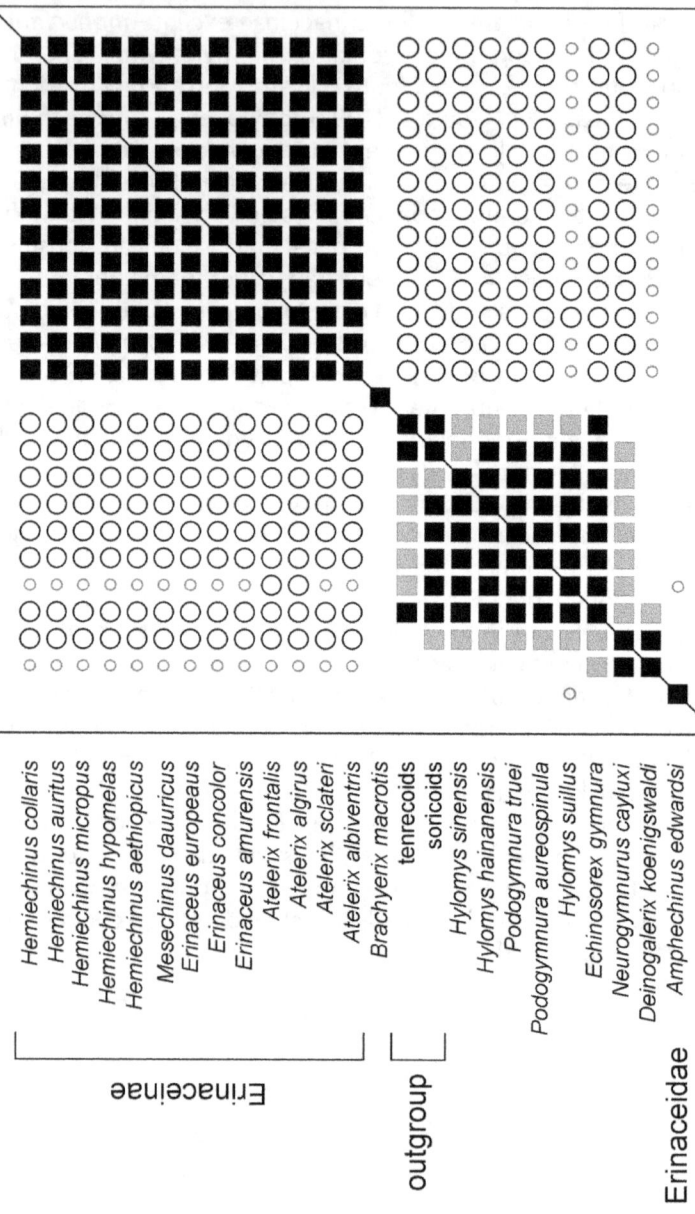

Figure 16. BDC bootstrap results for Erinaceidae, as calculated by BDISTMDS (relevance cutoff 0.9). Closed square indicate significant, positive BDC; open circles indicate significant, negative BDC. Black symbols indicate bootstrap values > 90% in a sample of 100 pseudoreplicates. Grey symbols represent bootstrap values ≤ 90%.

Animals 31

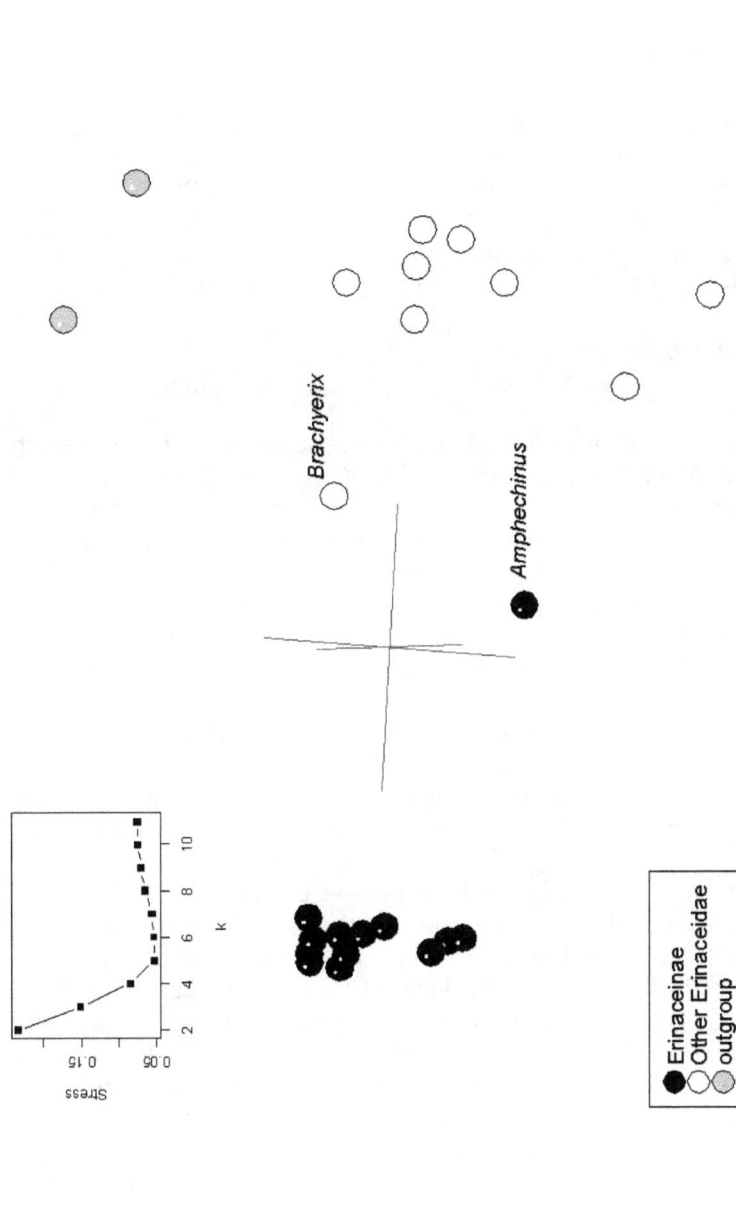

Figure 17. Three-dimensional MDS applied to Erinaceidae baraminic distances and the stress of $k$-dimensional MDS on the same baraminic distance matrix plotted as a function of the number of dimensions ($k$). Members of subfamily Erinaceinae are shown in black, other erinaceids in white, and outgroup taxa in gray.

### 2.8. Talpidae (Vertebrata: Mammalia: Erinaceomorpha)

Dataset published by Whidden (2000)

| | |
|---|---|
| Characters in published dataset: | 58 |
| Taxa in published dataset: | 13 |
| Character relevance cutoff: | 0.95 |
| Characters used to calculate BD: | 48 |
| Taxa used in BDC and MDS analysis: | 13 |
| Stress for 3D MDS: | 0.025 |
| $k_{min}$: | 3 |
| Median bootstrap value: | 93 |
| $F_{90}$: | 0.564 |

Moles have been classified into 31-42 species and 12-17 genera in the family Talpidae. McKenna and Bell (1997) list 45 genera, 30 of which are known only from the fossil record. North American moles are well known to yard owners for their subterranean activities, but other talpids, such as the Asian desmans, are known to be aquatic with webbed feet. Whereas Talpidae is generally recognized as a monophyletic group, the Soricidae and Erinaceidae have been suggested as possible sister taxa. Whidden (2000) generated a dataset of 58 myological characters and thirteen taxa to resolve the phylogeny of the Talpidae, particularly with respect to the Old World and New World fossorial forms, *Talpa*, *Scalopus*, *Scapanus*, and *Parascalops*. The thirteen taxa consist of eleven extant genera of talpids and two outgroups (soricid *Blarina* and erinaceid *Atelerix*). The myological characters include muscles from the whole body.

All taxa were included in the baraminic distance calculations, and the relevance cutoff omitted only ten characters (5-7, 15-16, 43, 48, 55-56). The BDC results show four groups of taxa that have no significant, positive BDC with any other taxa (Figure 18). The first group contains the outgroups and the talpid *Uropsilus*; the second *Galemys*, *Desmana*, and *Condylura*; the third *Urotrichus*, *Scaptonyx*, and *Neurotrichus*; and the fourth the fossorial forms *Talpa*, *Scalopus*, *Scapanus*, and *Parascalops*. The only significant, negative BDC occurs between the fossorial talpid group and *Galemys*, *Desmana*, *Uropsilus*, and the outgroups. The bootstrap values are also very high for the positive and negative correlations (median 93%).

The MDS results were extremely good, with a minimal stress of only 0.025 at three dimensions. The clustering pattern is quite diffuse,

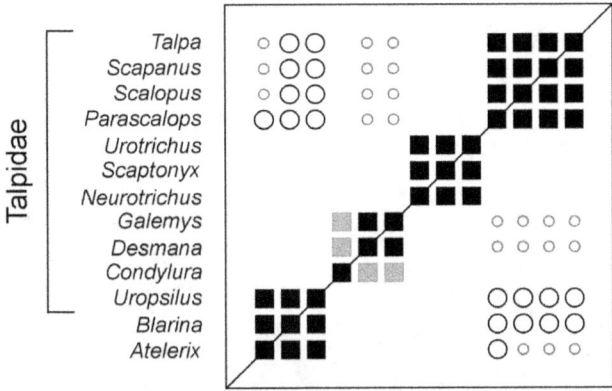

Figure 18. BDC bootstrap results for Talpidae, as calculated by BDISTMDS (relevance cutoff 0.95). Closed square indicate significant, positive BDC; open circles indicate significant, negative BDC. Black symbols indicate bootstrap values >90% in a sample of 100 pseudoreplicates. Grey symbols represent bootstrap values ≤90%.

forming roughly an arc with the fossorial talpids at one end and the outgroup/*Uropsilus* group at the other (Figure 19). The positions of *Neurotrichus*, *Urotrichus*, and *Scaptonyx* are nearly indistinguishable, and *Condylura* is slightly separated from a tight cluster of *Desmana* and *Galemys*.

Two groups found in the BDC and MDS results correspond to clades found by Whidden (2000): the fossorial talpids and the *Urotrichus/ Scaptonyx/Neurotrichus* group. The remaining two groups identified here are not monophyletic in Whidden's analysis but are closely allied. Whidden found that *Uropsilus* was the most basal of the talpids, which would be consistent with the observed clustering of *Uropsilus* with the outgroup taxa in BDC and MDS. Finally, *Desmana*, *Galemys*, and *Condylura* are paraphyletic in Whidden's phylogeny.

Whereas the BDC and MDS results correspond well to Whidden's phylogeny, they are difficult to interpret from a baraminological perspective. One possible interpretation is that the fossorial talpids alone correspond to a holobaramin due to the significant, negative BDC with other taxa in the Talpidae and outgroups; however, the semicircular pattern of taxa might account for the significant, negative BDC as was the case for negative correlation observed for the linear pattern of equid taxa (Cavanaugh et al. 2003). The positive BDC between *Uropsilus* and outgroup taxa from two different families is also somewhat surprising. It could suggest that Uropsilus should be omitted from any talpid baramin, but it could also merely indicate that the myological characters are not capable of distinguishing talpids from nontalpids. The baraminic status of Talpidae is therefore inconclusive.

# 34  Animal and Plant Baramins

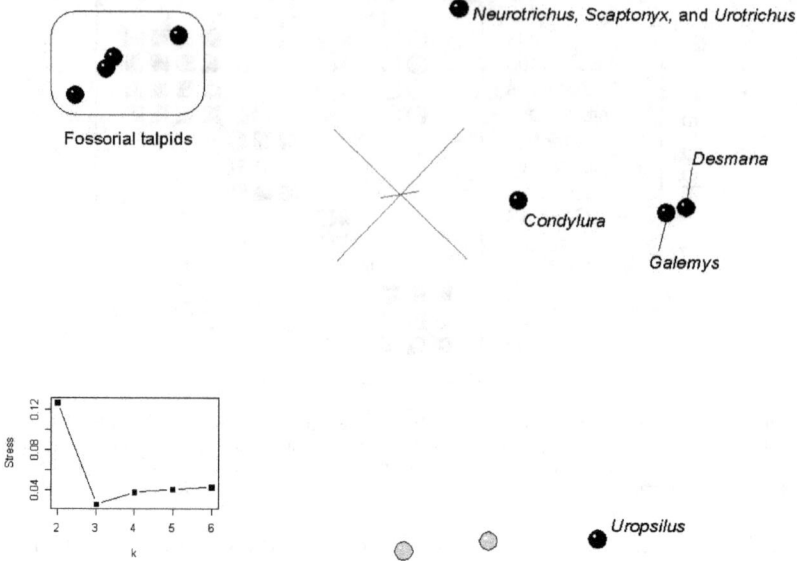

Figure 19. Three-dimensional MDS applied to Talpidae baraminic distances and the stress of *k*-dimensional MDS on the same baraminic distance matrix plotted as a function of the number of dimensions (*k*). Talpids are shown in black, outgroup taxa in gray.

## 2.9. Tenrecidae (Vertebrata: Mammalia: Erinaceomorpha)

Dataset published by Asher (1999)

| | |
|---|---|
| Characters in published dataset: | 71 |
| Taxa in published dataset: | 35 |
| Character relevance cutoff: | 0.9 |
| Characters used to calculate BD: | 54 |
| Taxa used in BDC and MDS analysis: | 35 |
| Stress for 3D MDS: | 0.236 |
| $k_{min}$: | 9 |
| Median bootstrap value: | 73 |
| $F_{90}$: | 0.143 |

The tenrecs are about 24 species of highly diverse shrew-like mammals restricted to Madagascar and central Africa. Their fossil record includes forms from east Africa and Madagascar. Three extant tenrec subfamilies are found only in Madagascar (Tenrecinae, Oryzorictinae, Geogalinae), while the extant African forms are classified in subfamily Potamogalinae. Asher (1999) compiled a dataset consisting of 71 characters and 35 taxa to evaluate Tenrecidae phylogeny. The characters include 35 cranial, 17 dental, and 19 postcranial characters. The taxa are highly diverse with all ten extant tenrecid genera and 25 outgroup taxa. The outgroups include marsupials, xenarthrans, carnivores, other members of the order Lipotyphla, which contains the talpids, erinaceids, chrysochlorids, and soricids in addition to the tenrecids.

All taxa were retained for baraminic distance calculation, and 17 characters were omitted at the character relevance cutoff of 0.9. The remaining characters consisted of 33 cranial, 17 dental, and four postcranial characters (characters omitted: 25, 30, 53-54, 56, 60-71). The BDC results show that the tenrecid taxa form a block of taxa united by significant, positive BDC (Figure 20). Eight tenrecids are also positively correlated with the extinct lipotyphlan *Apternodus*, and the small African water shrew (Tenrecidae: *Micropotamogale*) is correlated positively with the giant golden mole (Chrysochloridae: *Chrysospalax*). A large block of taxa apparent from the BDC results includes marsupials, hedgehogs, carnivores, and the aardvark (*Orycteropus*). The bootstrap values are poor, with only 14.3% of taxon pairs having bootstrap values >90%.

The MDS results are also somewhat poor, with a stress of 0.236 at three dimensions (Figure 21). The Tenrecidae form two clusters of taxa corresponding to the Tenrecinae and the remaining tenrecids.

*Apternodus* is adjacent to the non-tenrecine tenrecids, and the family Chrysochloridae is adjacent to both groups. The remaining taxa are dispersed widely with little obvious clustering.

It is unlikely that any reliable baraminic classifications can be inferred from these results. While there does seem to be some reasonable clustering within the Tenrecidae, for example separating the subfamily Tenrecinae from the rest of the tenrecids, the clustering pattern of the outgroup is quite puzzling. For example, one might imagine that the marsupials *Macropus* and *Didelphis* would be closest to each other, as measured by baraminic distances, but that is not the case. The baraminic distance between *Macropus* and *Didelphis* is 0.333, but *Didelphis* is actually closest to the carnivore *Nandinia* (baraminic distance 0.222). This is consistent with Asher's (1999) phylogenies, which do not show *Macropus* and *Didelphis* as sister taxa. Conservatively, the Tenrecidae could be classified as a monobaramin, but considering the positive correlation between carnivores and marsupials in the same dataset, even this conclusion is dubious.

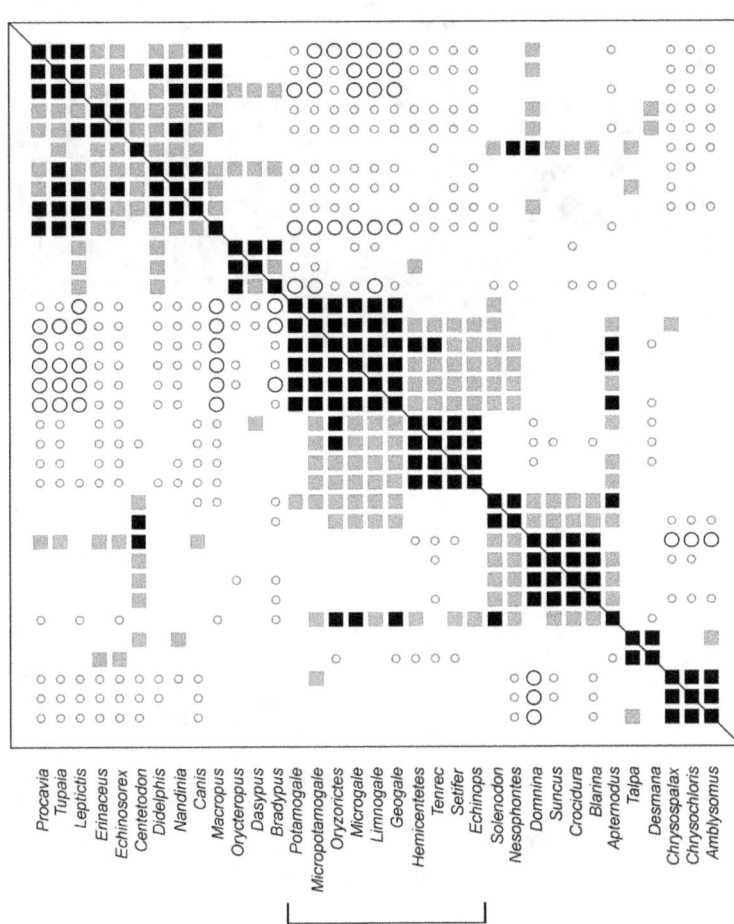

Figure 20. BDC bootstrap results for Tenrecidae, as calculated by BDISTMDS (relevance cutoff 0.9). Closed square indicate significant, positive BDC; open circles indicate significant, negative BDC. Black symbols indicate bootstrap values >90% in a sample of 100 pseudoreplicates. Grey symbols represent bootstrap values ≤90%.

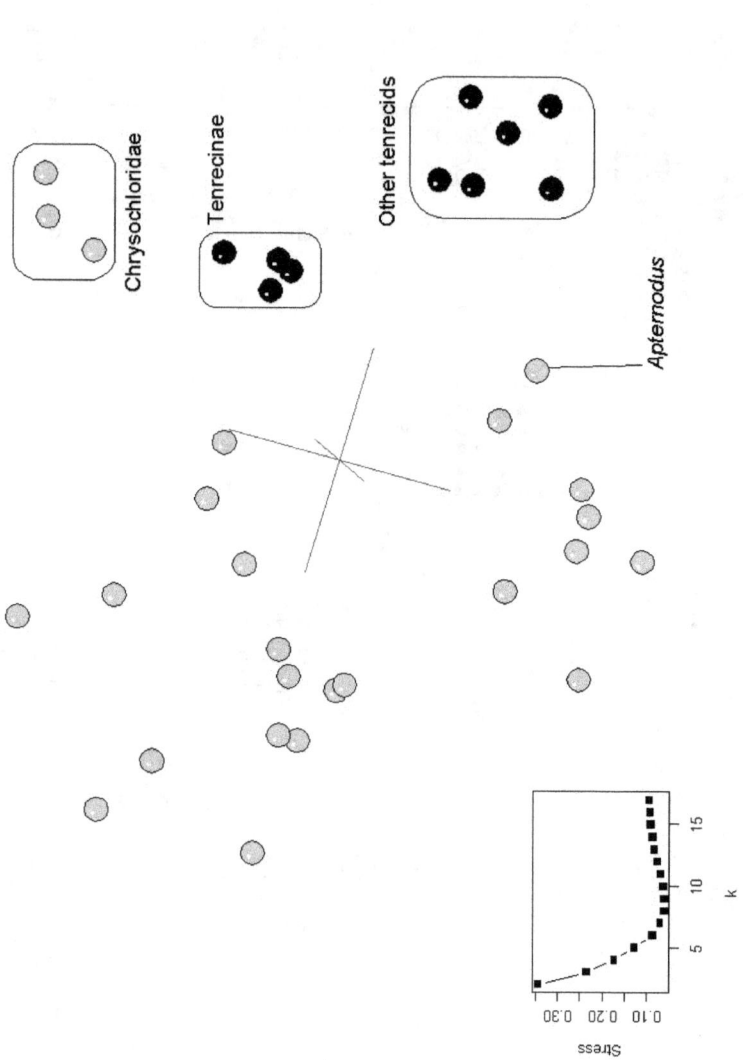

Figure 21. Three-dimensional MDS applied to Tenrecidae baraminic distances and the stress of $k$-dimensional MDS on the same baraminic distance matrix plotted as a function of the number of dimensions ($k$). Tenrecids are shown in black and outgroup taxa in gray.

## 2.10. Mormoopidae (Vertebrata: Mammalia: Chiroptera)

Dataset published by Simmons and Conway (2001)

| | |
|---|---|
| Characters in published dataset: | 209 |
| Taxa in published dataset: | 17 |
| Character relevance cutoff: | 0.95 |
| Characters used to calculate BD: | 106 |
| Taxa used in BDC and MDS analysis: | 16 |
| Stress for 3D MDS: | 0.169 |
| $k_{min}$: | 7 |
| Median bootstrap value: | 93 |
| $F_{90}$: | 0.533 |

The eight extant species of Neotropical spectacled bats are classified into two genera (*Mormoops* and *Pteronotus*) of the family Mormoopidae. Two Quaternary fossil species have also been assigned to the Mormoopidae. Phyllostomidae (see section 2.11) is thought to be the sister group to the mormoopids, and in some studies Mormoopidae appears to be paraphyletic with Phyllostomidae. Together with the Noctilionidae, the Phyllostomidae and Mormoopidae are classified in the superfamily Noctilionoidea, which is supported by mitochondrial DNA studies (see Simmons and Conway 2001). The morphological dataset of Simmons and Conway (2001) contains 209 characters scored for nine mormoopids and eight outgroup taxa. The characters consist of 27 cranial, 20 dental, 69 cranial soft tissue, 60 postcranial skeletal, and 33 postcranial soft tissue characters. The outgroup taxa are three phyllostomids (*Macrotus waterhousii*, *Macrotus californicus*, and *Artibeus jamaicensis*), two noctilionids (*Noctilio leporinus* and *Noctilio albiventris*), two mystacinids (*Mystacina tuberculata* and *Mystacina robusta*), and one emballonurid (*Saccopteryx bilineata*).

For calculating baraminic distances, the fossil species *Pteronotus pristinus* was omitted due to extremely low taxic relevance (15.8%). After filtering at a character relevance cutoff of 0.95 (meaning no unknown character states), 106 characters were retained for baraminic distance calculation. The remaining characters were 27 cranial, 20 dental, 27 cranial soft tissue, and 32 postcranial skeletal characters. Four groups of taxa are revealed in the BDC results corresponding to the four families represented by more than one taxon (Figure 22). All members of each group share significant, positive BDC with each other but not with any member of any other family. Bootstrapping results for the positive correlation is quite good, with most taxon pairs having >90% bootstrap

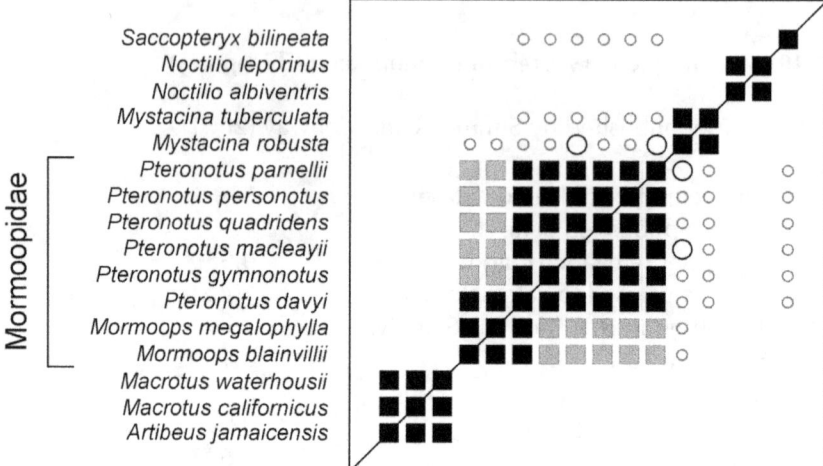

Figure 22. BDC bootstrap results for Mormoopidae, as calculated by BDISTMDS (relevance cutoff 0.95). Closed square indicate significant, positive BDC; open circles indicate significant, negative BDC. Black symbols indicate bootstrap values >90% in a sample of 100 pseudoreplicates. Grey symbols represent bootstrap values ≤90%.

values. Only intergeneric comparisons within the Mormoopidae had bootstrap values <90%. Significant, negative correlation is observed between the mystacinids and the mormoopids and between *Saccopteryx* and the mormoopids. None of the noctilionids or phyllostomids exhibited negative or positive correlation with any of the mormoopids.

MDS reveals an obvious gap between the cluster of mormoopids and all of the outgroup taxa, including the phyllostomids and the noctilionids (Figure 23). Within the Mormoopidae, the two genera *Mormoops* and *Pteronotus* are slightly separated but not to the same degree as the outgroup taxa. Stress values indicate that these MDS results are fairly typical, with a 3D stress of 0.17 and a minimal stress of 0.02 at seven dimensions.

These results are somewhat consistent with the phylogenies published by Simmons and Conway (2001). They found that all families formed monophyletic lineages, and that Phyllostomidae was the sister taxon to the Mormoopidae. Intriguingly, their results placed the Mystacinidae within the Noctilionoidea with Noctilionidae being the most basal lineage of that clade. In contrast, the significant, negative BDC results seem to indicate a clear separation between the mystacinids and mormoopids.

Taken together, the BDC and MDS results imply that the mormoopids are a holobaramin. The significant, positive BDC between all members of the Mormoopidae establish it as a monobaramin, and the significant, negative BDC with three of the outgroup taxa suggest a discontinuity

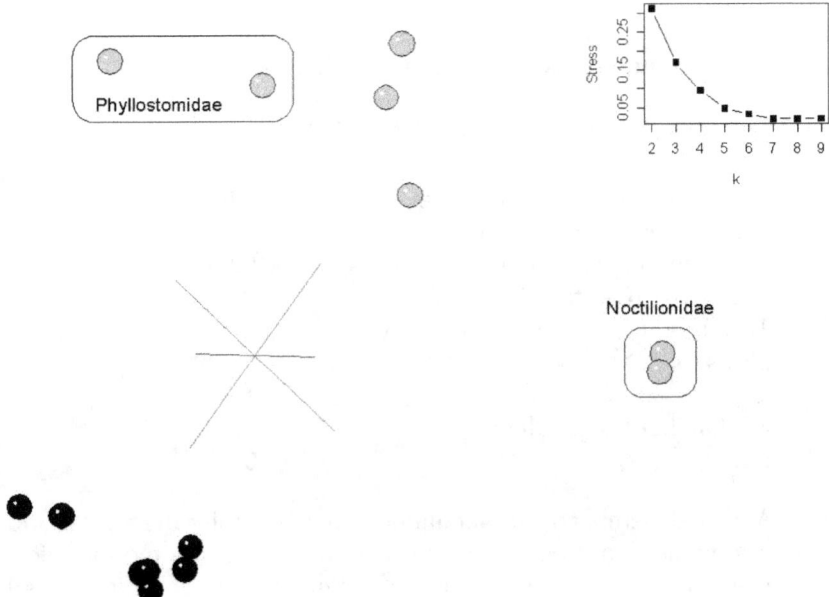

Figure 23. Three-dimensional MDS applied to Mormoopidae baraminic distances and the stress of k-dimensional MDS on the same baraminic distance matrix plotted as a function of the number of dimensions (k). Mormoopids are shown in black, outgroup taxa in gray.

around the family. The MDS results support the holobaraminic status of the Mormoopidae by revealing a gap between the mormoopids and the other noctilionoid outgroup taxa, which do not show significant, negative BDC with the mormoopids. While it is possible that Noctilionoidea could be a holobaramin and mormoopidae only a well-defined monobaramin within it, this holistic dataset of characters strongly imply that Mormoopidae is a holobaramin.

## 2.11. Phyllostomidae (Vertebrata: Mammalia: Chiroptera)

Dataset published by Wetterer et al. (2000)

| | |
|---|---|
| Characters in published dataset: | 136 |
| Taxa in published dataset: | 64 |
| Character relevance cutoff: | 0.95 |
| Characters used to calculate BD: | 90 |
| Taxa used in BDC and MDS analysis: | 57 |
| Stress for 3D MDS: | 0.228 |
| $k_{min}$: | 10 |
| Median bootstrap value: | 78 |
| $F_{90}$: | 0.333 |

A second member of the Noctilionoidea superfamily, the New World Phyllostomidae is a much more diverse group than the mormoopids. There are more than 140 species of phyllostomids classified in 49 different genera. Phyllostomid species also exhibit a remarkable variety of feeding habits, including sanguivory (e.g., *Desmodus*), frugivory (e.g., *Artibeus*), insecivory (e.g., *Lonchorhina*), and nectarivory (e.g., *Lionycteris*). McKenna and Bell (1997) classify the phyllostomids into four subfamilies, while other specialists have recognized as many as eight subfamilies (see Wetterer et al. 2000). Wetterer et al. (2000) examined phyllostomid phylogeny using 136 morphological characters and 14 molecular characters. The morphological characters consist of 35 craniodental, 16 postcranial skeletal, and 85 soft tissue characters. Wetterer et al. (2000) sampled 60 genera and species of phyllostomids and three noctilionoid outgroups (*Noctilio*, *Pteronotus*, and *Mormoops*).

For baraminic distance calculations, all molecular characters were omitted based on the experience of Wood (2002), and taxa with taxic relevance <0.66 were also omitted (*Artibeus* subgenus *Koopmania*, *Glyphonycteris*, *Lampronycteris*, *Musonycteris*, *Scleronycteris*, and *Vampyress abidens*). With the 57 remaining taxa and 136 characters, the character relevance cutoff of 0.95 retained 90 characters for calculating baraminic distances (characters 5-38, 42-67, 71-73, 84-89, and 107-127 were retained). BDC results show five groups of taxa that correspond roughly to the outgroup taxa and the four phyllostomid subfamilies (Desmodontinae, Glossophaginae, Phyllostominae, and Stenodermatinae) (Figure 24). Many taxa correlate positively with members of more than one subfamily, and the cavern leaf-nosed bat (Glossophaginae: *Brachyphylla*) and the genus *Rhinophylla*

(Stenodermatinae) both correlate positively with members of all four subfamilies. Within the phyllostomids, there is significant, negative BDC primarily between the glossophagines and stenodermatines. The outgroup taxa correlate positively with phyllostomines and negatively with glossophagines and stenodermatines. There is no significant BDC between the desmodontines and the outgroup taxa. Overall, bootstrap values are only moderately high (median 78%), but for taxa within each subfamily, bootstrap values are generally >90%.

The MDS results are poorer than those of the mormoopids (Figure 25). At three dimensions, the stress was 0.22, and the minimal stress was 0.04 at 10 dimensions. The taxa form a diffuse cluster with no separate clusters. Three subfamilies (Phyllostominae, Stenodermatinae, and Glossophaginae) form "lobes" that extend from the center of the main cluster. The desmodontines are adjacent to the Stenodermatinae, and the outgroup taxa are adjacent to the Phyllostominae. In the center of the cluster are taxa that correlate with 3-4 subfamilies (*Rhinophylla*, *Brachyphylla*, and *Carollia*).

Based on these results alone, there is no evidence of discontinuity separating the phyllostomids and noctilionoid outgroup taxa. The entire group of taxa appear to be a single monobaramin, which contrasts with the results of the previous study which suggested a discontinuity between Mormoopidae and nonmormoopids. One possible resolution is a holobaramin comprising the entire superfamily Noctilionoidea (Noctilionidae, Mormoopidae, and Phyllostomidae). The present datasets may be insufficient to resolve this question. Although the Mormoopidae dataset seems to indicate discontinuity between mormoopids and phyllostomids, this is only an inference based on the pattern of taxa in MDS. The Mormoopidae BDC results show no significant, negative BDC between mormoopids and phyllostomids. If the Mormoopidae dataset had a broader sampling of the Phyllostomidae, continuity between the two families may have been apparent. In this Phyllostomidae dataset, no outgroup taxa from outside the Noctilionoidea were utilized, making it impossible to tell if there is discontinuity around the superfamily. Future studies could easily rectify either of these problems and thereby clarify the baraminic status of the Noctilionoidea. Given the present data, however, it seems that the Mormoopidae should be tentatively identified as a holobaramin and the Phyllostomidae as a monobaramin only. Creationists ought to continue studying this family of bats, since the amazing variety of feeding habits in the Phyllostomidae would make an excellent model system to study the origin of carnivory as a part of post-Fall natural evil.

# Animal and Plant Baramins

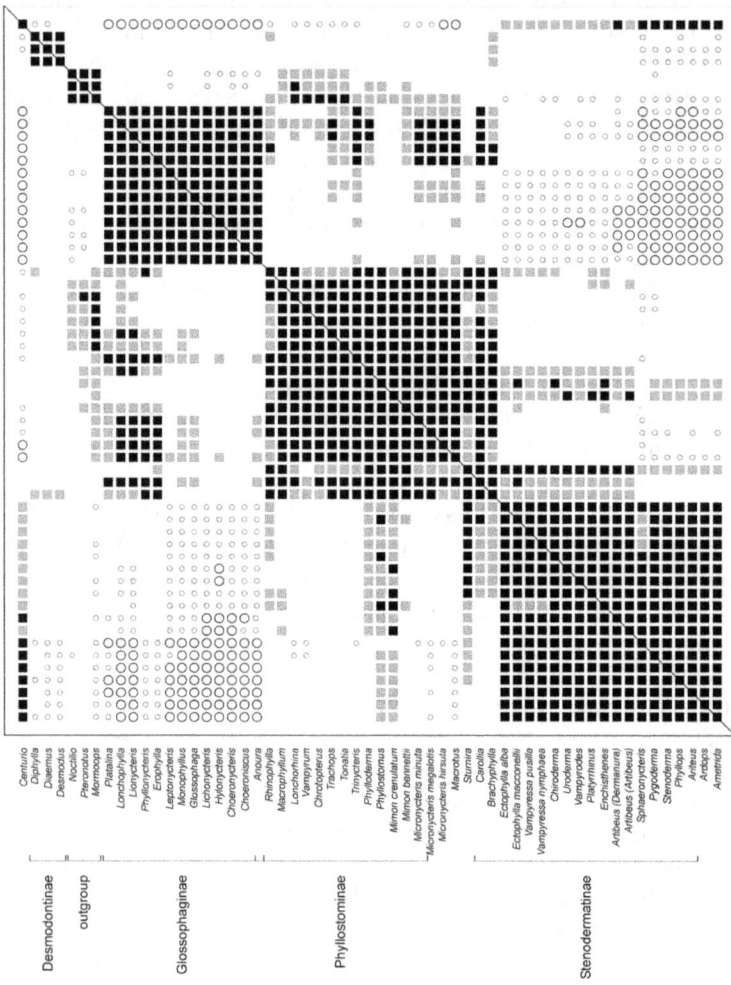

Figure 24. BDC bootstrap results for Phyllostomidae, as calculated by BDISTMDS (relevance cutoff 0.95). Closed square indicate significant, positive BDC; open circles indicate significant, negative BDC. Black symbols indicate bootstrap values >90% in a sample of 100 pseudoreplicates. Grey symbols represent bootstrap values ≤90%.

# Animals 45

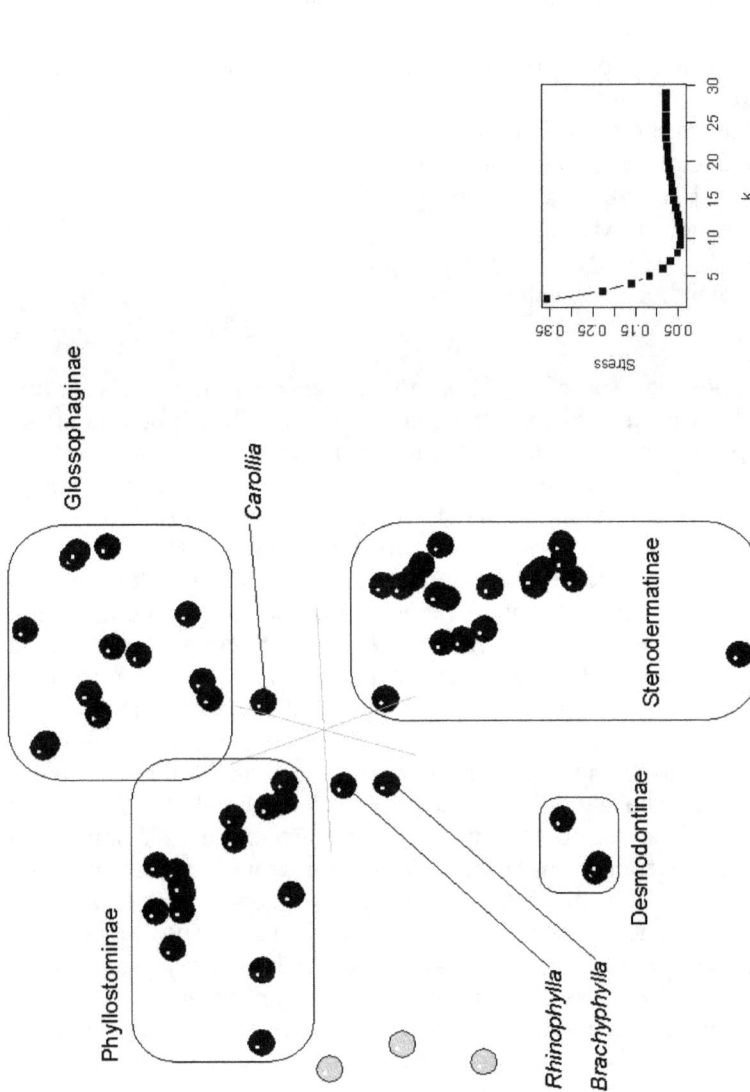

Figure 25. Three-dimensional MDS applied to Phyllostomidae baraminic distances and the stress of $k$-dimensional MDS on the same baraminic distance matrix plotted as a function of the number of dimensions ($k$). Phyllostomids are shown in black and outgroup taxa in gray.

## 2.12. Hippopotamidae (Vertebrata: Mammalia: Artiodactyla)

Dataset published by Boisserie et al. (2005)

| | |
|---|---|
| Characters in published dataset: | 80 |
| Taxa in published dataset: | 32 |
| Character relevance cutoff: | 0.9 |
| Characters used to calculate BD: | 62 |
| Taxa used in BDC and MDS analysis: | 30 |
| Stress for 3D MDS: | 0.193 |
| $k_{min}$: | 7 |
| Median bootstrap value: | 81 |
| $F_{90}$: | 0.34 |

Hippos are some of the most recognizable mammals in the world. There are two extant species of the family Hippopotamidae, the hippopotamus (*Hippopotamus amphibius*) and the pygmy hippopotamus (*Hexaprotodon sivalensis*). Fossil hippopotamids include the Miocene *Kenyapotamus*, the Miocene *Archaeopotamus*, the Pliocene *Trilobophorus*, the Pliocene *Saotherium*, and the Pleistocene *Choeropsis*. Fossil hippos are known from Europe, Asia, and Africa. Closely allied with the hippos are the Anthracotheriidae, a family of large ungulates known only from the fossil record of Asia, North America, and some Mediterranean islands (Mallorca and Balearic). Recently, numerous molecular studies suggest that the hippos are the sister taxon to the whales.

To resolve the relationships of the fossil anthracotheres, hippos, and whales, Boisserie et al. (2005) assembled a dataset of 80 characters scored for 32 taxa. The characters are 72 craniodental and eight postcranial characters. Six major groups of mammals are represented in the sampled taxa: Tayassuidae, Suidae, Hippopotamidae, Anthracotheriidae, Archaeoceti, and ruminants. The hippos are represented by six taxa: *Kenyapotamus*, *Choeropsis*, *Saotherium*, *Archaeopotamus*, *Hexaprotodon*, and *Hippopotamus*. Eight anthracotheres are present: *Libycosaurus*, *Merycopotamus*, *Elomeryx*, *Brachyodus*, *Anthracotherium*, *Microbunodon*, *Anthracokeryx*, and *Siamotherium*. Whales are represented by the archaeocetes *Pakicetus* and *Artiocetus*. The phylogeny of Boisserie et al. (2005) nested the hippos within a paraphyletic Anthracotheriidae, both of which are sister taxa to the archaeocetes.

For baraminic distances, two taxa were omitted from the calculations due to low taxic relevance: the suoid outgroup *Xenohyus* (0.36 taxic relevance) and *Kenyapotamus* (0.26 taxic relevance). Using a character relevance cutoff of 0.9, 62 characters were retained for baraminic distance calculation, only one of which was postcranial. The BDC results show two blocks of taxa: one consisting of all hippopotamids and two anthracotheres, *Merycopotamus* and *Libycosaurus*, and the other consisting of all other taxa (Figure 26). Overall the bootstrap values are only moderate (median 81%) but within the hippopotamids + *Merycopotamus*/*Libycosaurus* all taxa share significant, positive BDC with >90% bootstrap values. Taxa from different groups usually have significant, negative BDC, with five exceptions. *Merycopotamus* and *Libycosaurus* have significant, positive BDC with the anthracotheres *Elomeryx* and *Brachyodus*, and *Archaeopotamus* is positively correlated with *Brachyodus*.

The MDS results show the same two clusters of taxa detected in the BDC analysis (Figure 27). The Anthracotheriidae are spread throughout both groups, but there is a definite gap between *Merycopotamus*/*Libycosaurus* and the remaining anthracotheres. *Brachyodus* and *Elomeryx* most closely approach the Hippopotamidae + *Merycopotamus*/*Libycosaurus*, which likely accounts for the correlation observed in the BDC results.

The baraminological interpretation of these results is straightforward: Hippopotamidae + *Merycopotamus*/*Libycosaurus* is a holobaramin. These taxa are united by significant, positive BDC and separated from most other taxa by significant, negative BDC. The MDS results corroborate the BDC and suggest that the adjacency of *Elomeryx* and *Brachyodus* accounts for the spurious positive correlation with members of the hippopotamid holobaramin.

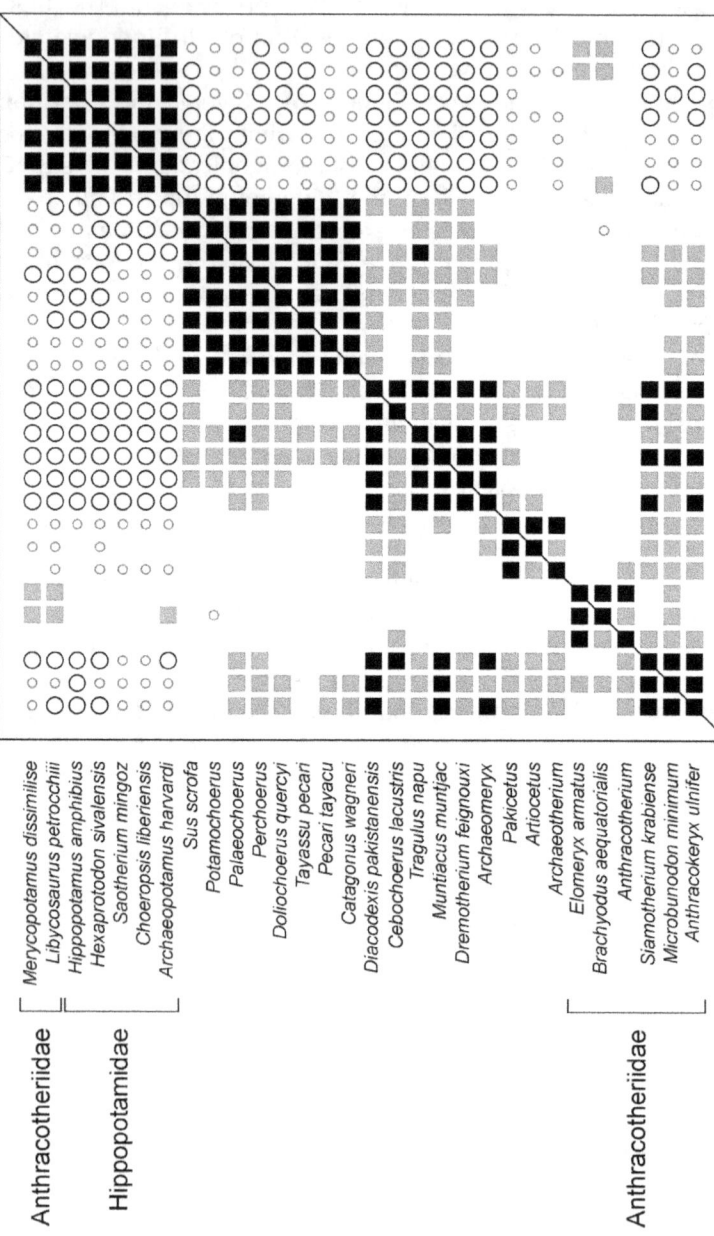

Figure 26. BDC bootstrap results for Hippopotamidae, as calculated by BDISTMDS (relevance cutoff 0.9). Closed square indicate significant, positive BDC; open circles indicate significant, negative BDC. Black symbols indicate bootstrap values > 90% in a sample of 100 pseudoreplicates. Grey symbols represent bootstrap values ≤ 90%.

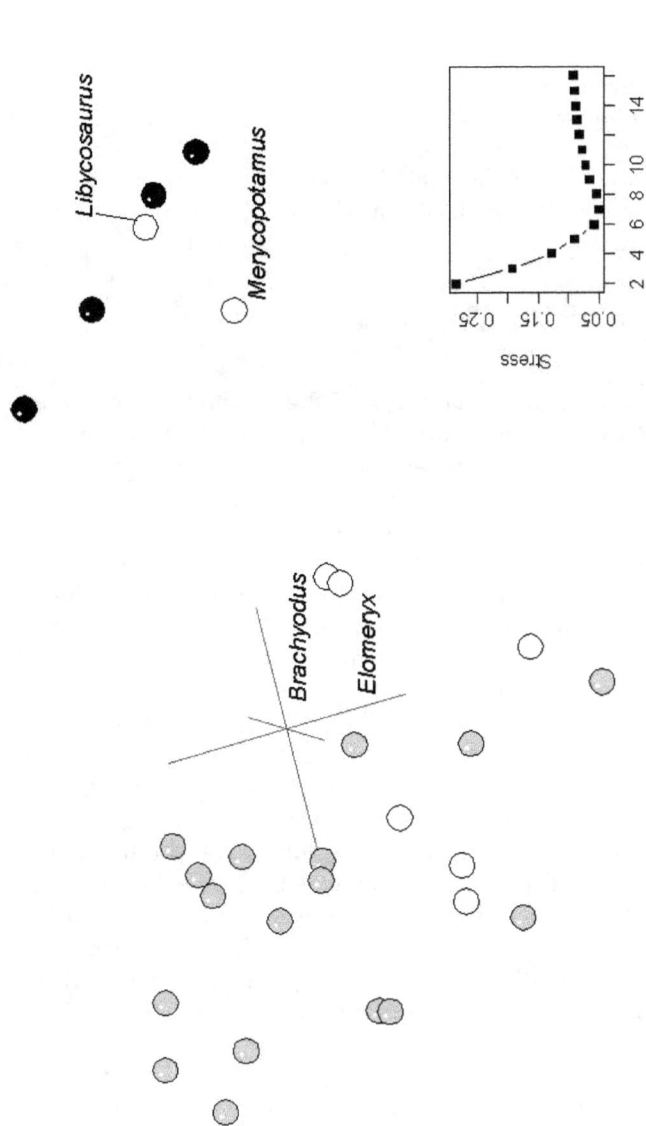

Figure 27. Three-dimensional MDS applied to Hippopotamidae baraminic distances and the stress of *k*-dimensional MDS on the same baraminic distance matrix plotted as a function of the number of dimensions (*k*). Hippopotamids are shown in black, anthracotheres in white, and outgroup taxa in gray.

## 2.13. Brontotheriidae (Vertebrata: Mammalia: Perissodactyla)

Dataset published by Mihlbachler et al. (2004)

| | |
|---|---|
| Characters in published dataset: | 40 |
| Taxa in published dataset: | 18 |
| Character relevance cutoff: | 0.9 |
| Characters used to calculate BD: | 31 |
| Taxa used in BDC and MDS analysis: | 14 |
| Stress for 3D MDS: | 0.135 |
| $k_{min}$: | 5 |
| Median bootstrap value: | 77 |
| $F_{90}$: | 0.187 |

With their elaborate nasal horns, the brontotheres are among the most well-known extinct mammal groups. According McKenna and Bell (1997), there are 43 genera of brontotheres in eight subfamilies. Brontothere fossils are known from Eocene and Oligocene sediments in Asia, Europe, and North America. Mihlbachler et al.'s (2004) dataset consists of 40 characters and 18 taxa, including their newly-described *Aktautitan hippopotamus*. The taxa selection focuses exclusively on Asian and European taxa of subfamilies Embolotheriinae, Brontopinae, and Telmatheriinae. No non-brontothere outgroups were included. The characters are entirely craniodental.

At a relevance cutoff of 0.9, 31 characters were retained for baraminic distance calculation (1, 3-22, 24-31, 33, 40). Correlation results reveal sporadic correlation with relatively poor bootstrap values (Figure 28). Nearly all congeners displayed significant, positive BDC and high (>90%) bootstrap values. *Aktautitan* was positively correlated with both *Metatitan* species, and significant, positive BDC was observed between *Protitan, Palaeosyops, Epintanteoceras, Dolichorhinoides*, and *Rhinotitan*. No other intergeneric positive BDC was observed. Some significant, negative BDC was observed, primarily in comparisons with *Embolotherium* species. *Parabrontops* did not have any significant BDC with any other taxa.

The distribution of taxa in 3D MDS is diffuse with little clustering (Figure 29). Species of the same genus tend to be closest to each other, except for *Embolotherium grangeri*, which is slightly closer to *Parabrontops* than to *Embolotherium andrewsi*. The MDS results are otherwise unremarkable; the 3D stress and $k_{min}$ indicate a good fit between the baraminic distances and the MDS distances.

Baraminologically, this dataset is probably the poorest of all in this study for resolving baraminic relationships. The few craniodental characters are not holistic, and the focus on only three of the eight subfamilies prevents detecting discontinuity at the family level. Despite these drawbacks, the lack of any obvious discontinuity in this group of taxa would be consistent with a holobaramin consisting of more than just these three subfamilies. The taxa in this dataset can be tentatively assigned to the same monobaramin, pending future studies of more taxa and characters.

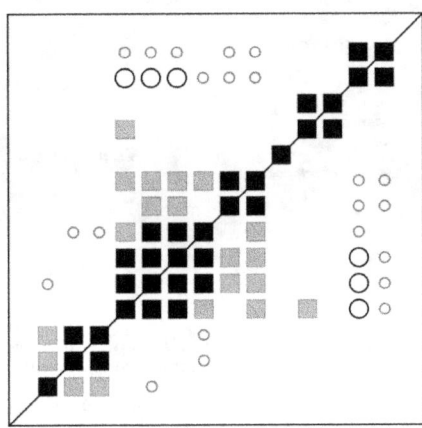

Figure 28. BDC bootstrap results for Brontotheriidae, as calculated by BDISTMDS (relevance cutoff 0.9). Closed square indicate significant, positive BDC; open circles indicate significant, negative BDC. Black symbols indicate bootstrap values >90% in a sample of 100 pseudoreplicates. Grey symbols represent bootstrap values ≤90%.

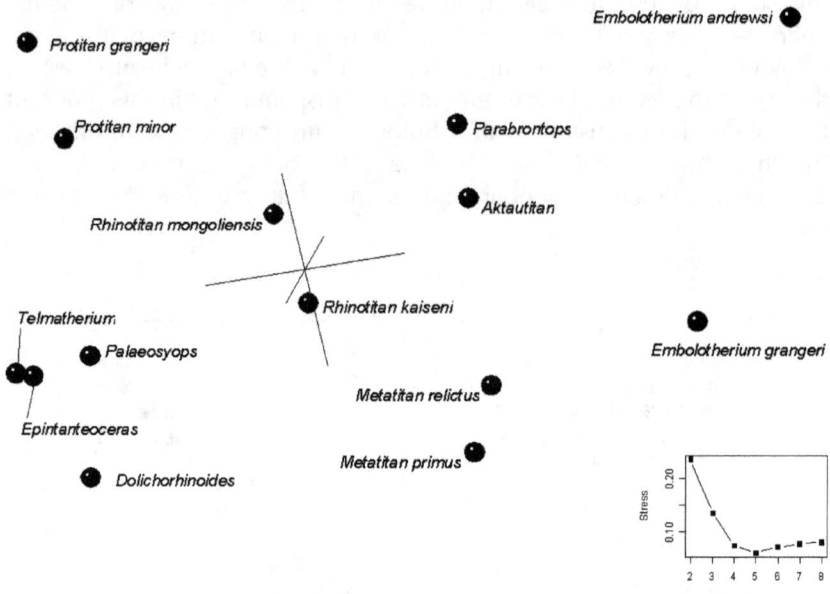

Figure 29. Three-dimensional MDS applied to Brontotheriidae baraminic distances and the stress of $k$-dimensional MDS on the same baraminic distance matrix plotted as a function of the number of dimensions ($k$).

## 2.14. Rhinocerotidae (Vertebrata: Mammalia: Perissodactyla)

Dataset published by Cerdeño (1995)

| | |
|---|---|
| Characters in published dataset: | 72 |
| Taxa in published dataset: | 46 |
| Character relevance cutoff: | 0.9 |
| Characters used to calculate BD: | 31 |
| Taxa used in BDC and MDS analysis: | 41 |
| Stress for 3D MDS: | 0.214 |
| $k_{min}$: | 7 |
| Median bootstrap value: | 65 |
| $F_{90}$: | 0.134 |

Rhinos are another recognizable group of perissodactyls, with their unique nasal horns consisting largely of fibrous keratin. Five extant species of rhinos are placed in four genera, but the rhino fossil record is rich: 68 genera are known from Asia, Europe, Africa, and North America. Based on a cladistic analysis, Cerdeño (1995) identified two subfamilies, Rhinocerotinae and Aceratheriinae. Taxa from both subfamilies are found in the Old and New Worlds. Cerdeño's dataset consists of 20 craniomandibular, 26 dental, and 26 postcranial characters, 72 in all. Taxa consist of 43 rhinos and three outgroup taxa of the family Hyracodontidae.

For baraminic distance calculations, five taxa with taxic relevance <0.66 were omitted: *Ninxiatherium* (0.46 taxic relevance), *Punjabitherium* (0.56 taxic relevance), *Mesaceratherium* (0.57 taxic relevance), *Iranotherium* (0.63 taxic relevance), and FAM 95544 (0.44 taxic relevance). For the remaining taxa, a 0.9 character relevance cutoff resulted in 31 of the 72 characters retained for baraminic distance calculation. The remaining characters consisted of nine craniomandibular, nineteen dental, and three postcranial characters.

Three groups are apparent from the BDC results (Figure 30). The first two groups correspond to the subfamilies Aceratheriinae and Rhinocerotinae, while the third group includes the remaining rhinos and outgroup taxa. Significant, positive BDC is most frequent within the groups, but taxa such as *Dicerorhinus* and *Gaindatherium* correlate positively with members of all three groups. The hyracodontids correlate positively with the rhinos of the third group and one member of the Aceratheriinae (*Amphicaenopsis*). There is no positive BDC between any rhinocerotine and a hyracodontid. Significant, negative BDC occurs primarily in comparisons between taxa of the Rhinocerotinae and the hyracodontid outgroups. Bootstrap values were unusually poor (median 65%).

MDS does not support three separate clusters of taxa (Figure 31). Instead, the taxa form a crudely triangular shape, with the hyracodontids adjacent to one of the triangular vertices. The Eocene *Teletaceras* is adjacent to the outgroup taxa.

The significant, positive BDC implies that the rhinos are members of a single monobaramin, but positive BDC with hyracodontids would suggest that they too are members of the same monobaramin. Though there is negative BDC, it does not separate any of the groups. The MDS results support the monobaraminic status of the rhinos but suggest that there might be some separation between the outgroups and the rhinos, although it is not statistically significant. It is possible that the hyracodontids are cobaraminic with rhinos, but several factors warrant caution in such a conclusion. First, as is the case with so many other datasets in this study, the characters are predominantly craniodental and therefore not holistic. Second, two of the outgroups are composite taxa representing subfamilies (Hyracodontinae and Eggysodontinae). Since there is no analysis at present to support the monobaraminic status of either of these subfamilies, any conclusions regarding their collective continuity or discontinuity with other taxa should be considered extremely tentative. Third, McKenna and Bell (1997) list fifteen hyracodontid genera beside the three taxa included in this dataset. It is possible that a broader sample of hyracodontid taxa might reveal evidence of discontinuity between hyracodontids and rhinocerotids. Nevertheless, based on the present data, there is no evidence of discontinuity between Rhinocerotidae and Hyracodontidae.

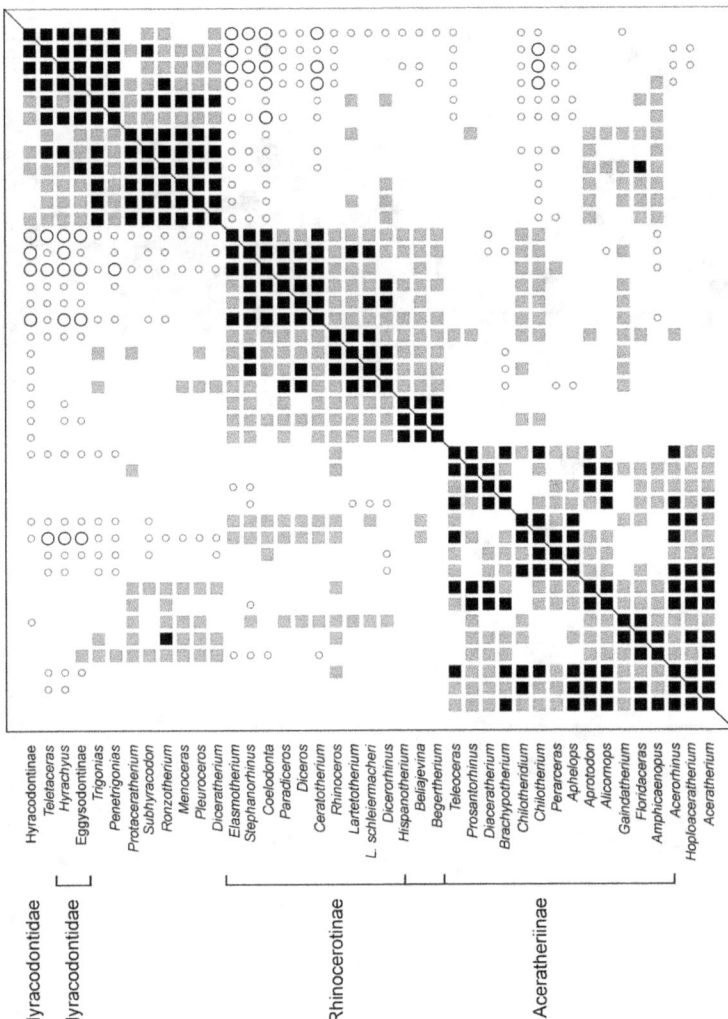

Figure 30. BDC bootstrap results for Rhinocerotidae, as calculated by BDISTMDS (relevance cutoff 0.9). Closed square indicate significant, positive BDC; open circles indicate significant, negative BDC. Black symbols indicate bootstrap values >90% in a sample of 100 pseudoreplicates. Grey symbols represent bootstrap values ≤90%.

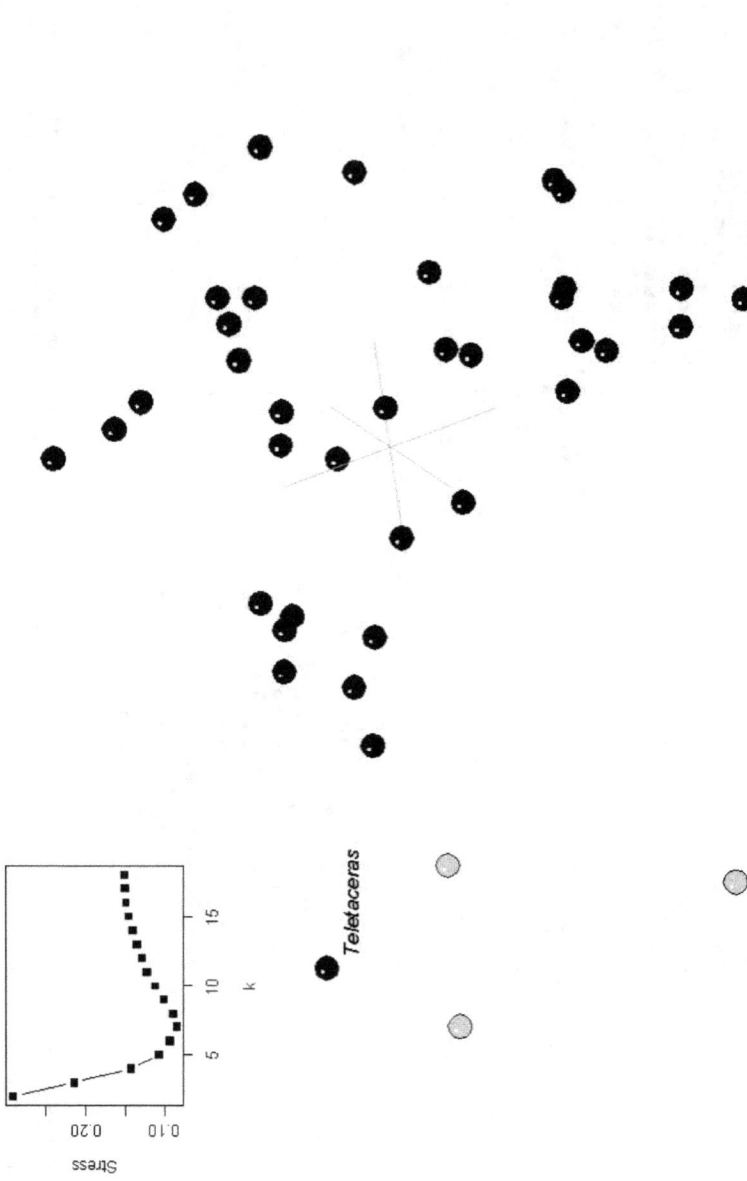

Figure 31. Three-dimensional MDS applied to Rhinocerotidae baraminic distances and the stress of *k*-dimensional MDS on the same baraminic distance matrix plotted as a function of the number of dimensions (*k*). Rhinocerotids are shown in black, outgroup taxa in gray.

## 2.15. Anserinae (Vertebrata: Aves: Anseriformes)

Dataset published by Livezey (1996)

| | |
|---|---|
| Characters in published dataset: | 175 |
| Taxa in published dataset: | 38 |
| Character relevance cutoff: | 0.95 |
| Characters used to calculate BD: | 160 |
| Taxa used in BDC and MDS analysis: | 32 |
| Stress for 3D MDS: | 0.181 |
| $k_{min}$: | 11 |
| Median bootstrap value: | 81 |
| $F_{90}$: | 0.282 |

The swans and geese are classified as a subfamily of the Anatidae, which also include the ducks. The entire family contains more than 400 species, primarily found in the northern hemisphere. The anatids have been analyzed previously within the basic type biology framework by Scherer (1993a), who recognized the entire family as a basic type (monobaramin) based on extensive hybridization. Of particular interest here are reported hybrids between geese (subfamily Anserinae) and ducks (subfamily Anatinae), which indicate that the two subfamilies belong to the same monobaramin. Livezey's (1996) analysis of the phylogeny of Anserinae was based on a data set of 175 characters and 39 taxa. The characters were composed of 15 cranial, 69 postcranial, 81 integumentary, and 10 ecomorphological characters. The taxa included a hypothetical anatid ancestor, six other outgroup taxa, and 32 anserines.

For baraminic distance calculations, taxa with taxic relevance less than 0.5 were omitted from the dataset: *Branta hylobadistes* (0.446 taxic relevance), *Chelychelynechen quassus* (0.154 taxic relevance), *Cnemiornis* (0.389 taxic relevance), *Geochen rhuax* (0.446 taxic relevance), *Ptaiochen pau* (0.229 taxic relevance), *Thambetochen chauliodous* (0.434 taxic relevance), and *Thambetochen xanion* (0.246 taxic relevance). Each of these are known only as fossils or subfossils and therefore have unknown integumentary character states. At a character relevance cutoff of 0.95, the remaining 32 taxa retained 160 of 175 characters for baraminic distance calculations.

BDC results reveal three very well-defined groups, corresponding to the tribe Cygnini (*Cygnus* and *Coscoroba*), tribe Anserini + *Cereopsis*, and the outgroup taxa (Figure 32). Within each group nearly all taxon pairs share significant, positive BDC, the only exception being nonsignificant

BDC between the hypothetical ancestor and Tadorninae. Most taxa in the Cygnini are negatively correlated with members of the Anserini and outgroups, and the hypothetical ancestor is negatively correlated with each member of Anserini but not Cygnini. Bootstrap values are reasonably high for the positive BDC but <90% for the negative BDC.

MDS reveals a striking pattern of taxa in the form of an irregular tetrahedron (Figure 33). At each vertex of the tetrahedron are the Cygnini, *Anser*, *Branta*, and the outgroups. In the center of the tetrahedron is the Australian Cape Barren goose *Cereopsis*. The taxa at each vertex are tightly clustered. The 3D stress is 0.18, indicating that the tetrahedron is probably a reasonable approximation of the distribution of these taxa given these character states.

Superficially, these results would seem to indicate discontinuity separating the Anserini and Cygnini, but several factors mitigate against such a conclusion. First and most important, the tetrahedral shape of the taxa in the MDS results indicates that any negative correlation in the BDC results is likely an artifact of the unusual geometry of the taxa. In other words, the characters appear to be diagnostic of the genera in the dataset and therefore reveal little about intergeneric relationships. Such character states would result in genera that are roughly equidistant, resulting in a tetrahedral shape in MDS. Tetrahedral geometry in MDS has been observed previously in the Sulidae (Wood 2005, pp. 133-146), although in that case, very little significant intergeneric BDC, positive or negative, was observed. Second, numerous *Anser/Cygnus* hybrids listed by Gray (1958) would warrant inclusion of the Anserini and Cygnini in the same monobaramin without statistical analysis. Considering the peculiar geometry observed in the MDS results, no reliable baraminic conclusions can be drawn from this dataset.

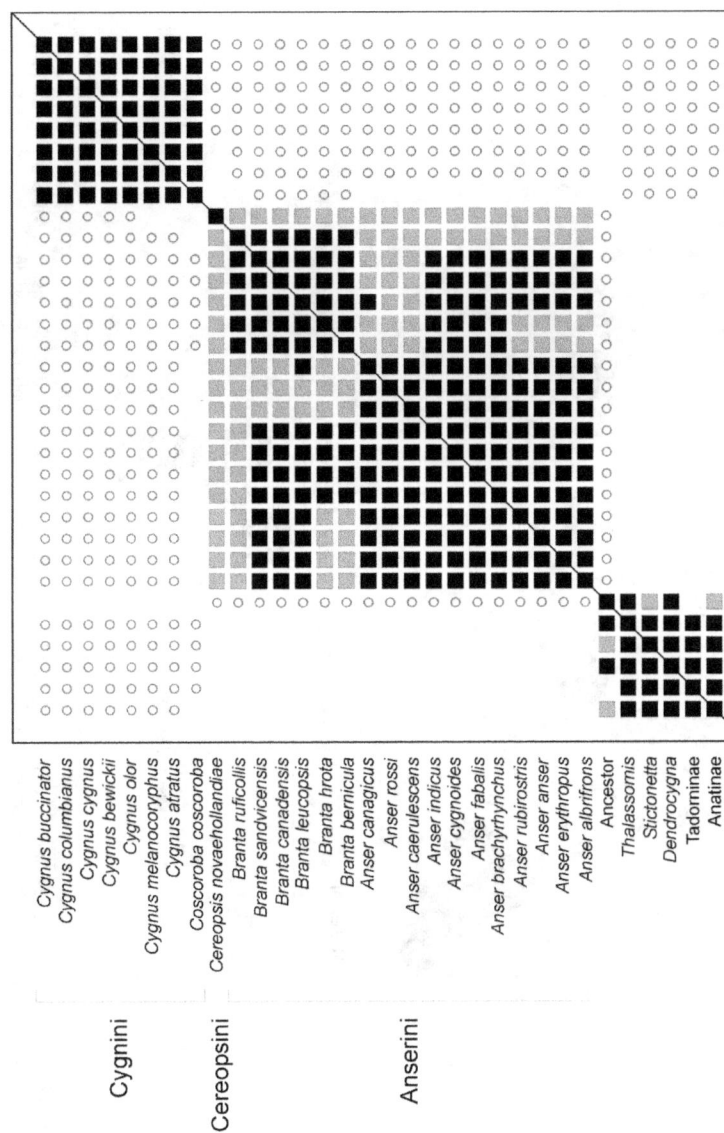

Figure 32. BDC bootstrap results for Anserinae, as calculated by BDISTMDS (relevance cutoff 0.95). Closed square indicate significant, positive BDC; open circles indicate significant, negative BDC. Black symbols indicate bootstrap values >90% in a sample of 100 pseudoreplicates. Grey symbols represent bootstrap values ≤90%.

60 Animal and Plant Baramins

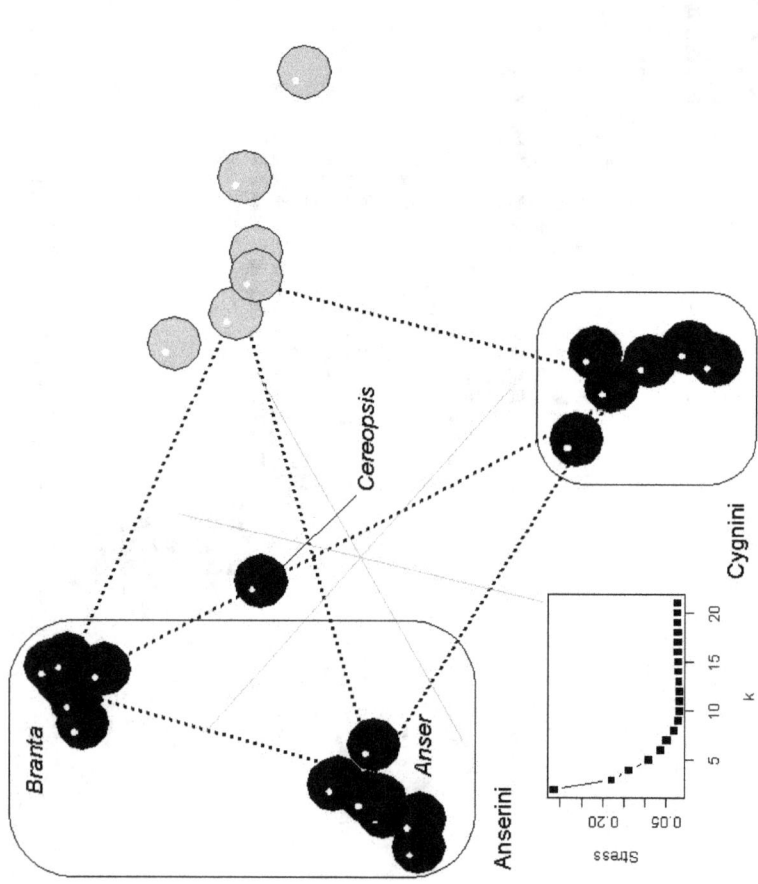

Figure 33. Three-dimensional MDS applied to Anserinae baraminic distances and the stress of *k*-dimensional MDS on the same baraminic distance matrix plotted as a function of the number of dimensions (*k*). Members of subfamily Anserinae are shown in black, outgroup taxa in gray.

## 2.16. Ardeidae (Vertebrata: Aves: Ciconiiformes)

Dataset published by Payne & Risley (1976)

| | |
|---|---|
| Characters in published dataset: | 33 |
| Taxa in published dataset: | 63 |
| Character relevance cutoff: | 0.95 |
| Characters used to calculate BD: | 33 |
| Taxa used in BDC and MDS analysis: | 59 |
| Stress for 3D MDS: | 0.168 |
| $k_{min}$: | 6 |
| Median bootstrap value: | 84 |
| $F_{90}$: | 0.433 |

The herons consist of approximately 62 species of wading birds in 15 genera, classified in the avian order Ciconiiformes. Herons are found worldwide, including on isolated islands, and are known to travel great distances (e.g., from Africa to South America, see Crosby 1972). There are currently three ardeid subfamilies recognized: Tigrisomatinae, the tiger herons; Botaurinae, the bitterns; and Ardeinae, the herons. Payne and Risley's (1976) analysis of heron phylogeny was based on 33 characters and 63 taxa. The characters were cranial (13) and postcranial (20), and the taxa consisted of 10 outgroups and 53 herons. The outgroups come primarily from other ciconiiform waders, mainly the ibises (Threskiornithidae) and storks (Ciconiidae).

For baraminic distance calculations, four herons were omitted from the dataset due to low taxic relevance: *Ardea imperialis* (0.61 taxic relevance), *Egretta eulophotes* (0.49 taxic relevance), *Nycticorax leuconotus* (0.27 taxic relevance), and *Zonerodius heliosylus* (0.27 taxic relevance). For the remaining 59 taxa, all 33 characters were retained after the 0.95 character relevance filter.

BDC results reveal two well-defined groups (Figure 34). The first group corresponds to the bittern subfamily Botaurinae. All botaurine taxon pairs share significant, positive BDC. The second group includes the remaining herons and the outgroup taxa. Significant, positive BDC is not observed for all taxon pairs of the second group, and several taxon pairs of the second group exhibit significant, negative BDC. Between the two groups, signficant, negative BDC is often observed, especially between the outgroup taxa and the botaurines. One botaurine, the zig-zag heron *Zebrilus undulatus* correlates positively with the three species of the ardeine genus *Nycticorax*, the night herons. Bootstrap values are quite good, with 43% of the taxon pairs having a >90% bootstrap value.

MDS reveals a tight but elongated cluster of Botaurinae with *Zebrilus* notably separated from the other botaurines and a much more diffuse cluster of Ardeinae, Tigrisomatinae, and outgroups (Figure 35). The most remote outgroup taxon is the shoebill, *Balaeniceps rex*, but the other outgroups cluster closely near the Ardeinae and Tigrisomatinae. The closest taxon to the shoebill is the boatbilled heron *Cochlearius*, which itself is separated from the main core of Ardeinae and Tigrisomatinae. Between *Zebrilus* and the bulk of the Ardeinae/Tigrisomatinae are the *Nycticorax* species.

Baraminologically, one possible interpretation of these results is that the botaurines are a holobaramin, discontinuous from the remaining ardeids. While such a small holobaramin seems unlikely given results in other groups that place the holobaramin at the rank of family or higher, there is little from this dataset to warrant against such a conclusion. The significant, positive BDC between *Zebrilus* and *Nycticorax* could be explained as spurious due to their adjacency but not proximity. Significant, negative BDC between botaurines and ardeines/tigrisomatines, is nevertheless sporadic and all with bootstrap values <90%. Only one botaurine, *Zebrilus*, is negatively correlated with most of the ardeines and tigrisomatines.

Another possible interpretation is that the entire set of taxa, including the outgroups are members of a single monobaramin. Since most of the taxa are ciconiiforms, it would suggest that the entire order Ciconiiformes is a holobaramin. The problem with this interpretation is the presence of non-ciconiiform outgroups (shoebill and flamingo) that correlate positively with several ardeine taxa. The flamingo (*Phoenicopterus*) correlates positively with almost all of the ardeines, and the shoebill (*Balaeniceps*) correlates positively with *Nycticorax* and *Cochlearius*. Unless we are willing to broaden the holobaramin beyond just one order, it would seem the more conservative interpretation would be a Botaurinae holobaramin.

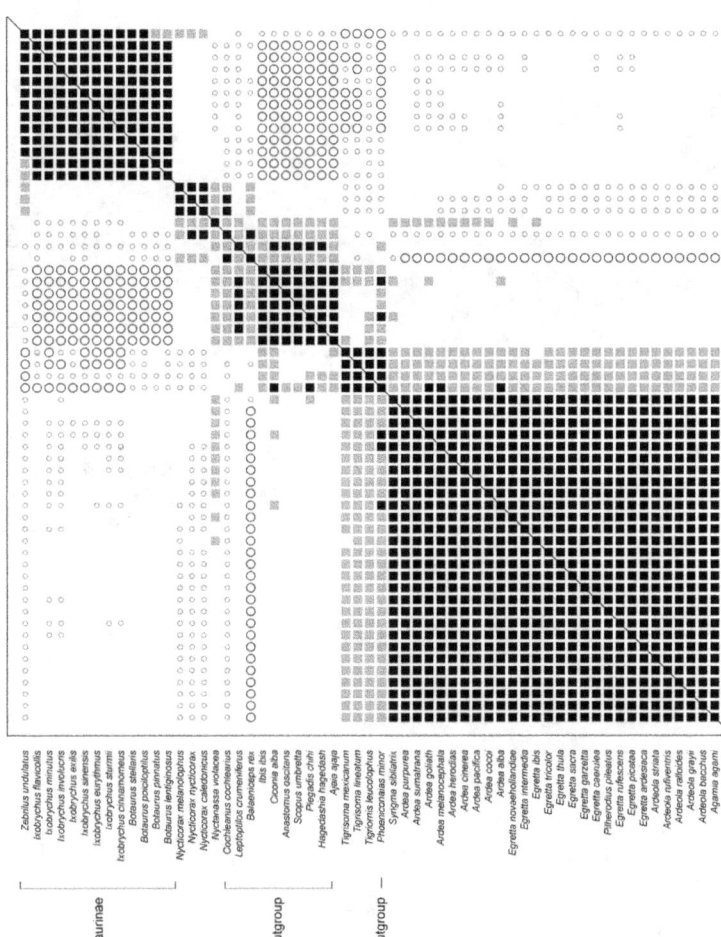

Figure 34. BDC bootstrap results for Ardeidae, as calculated by BDISTMDS (relevance cutoff 0.95). Closed square indicate significant, positive BDC; open circles indicate significant, negative BDC. Black symbols indicate bootstrap values > 90% in a sample of 100 pseudoreplicates. Grey symbols represent bootstrap values ≤ 90%.

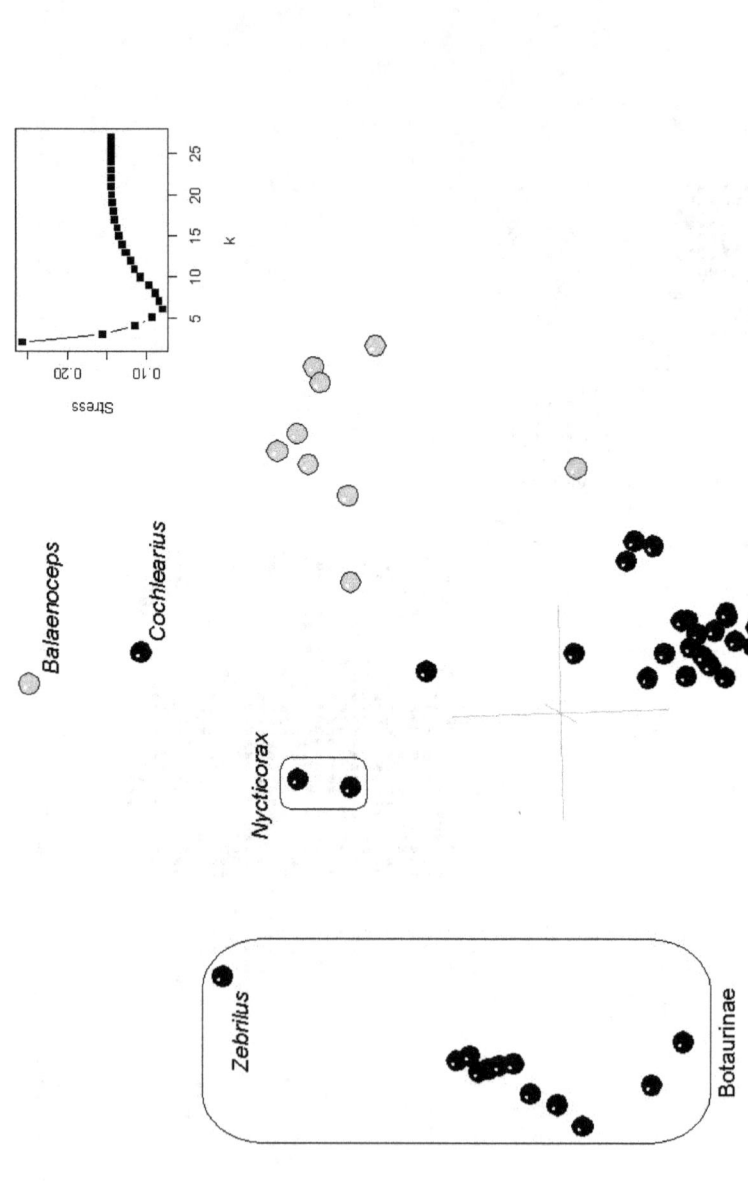

Figure 35. Three-dimensional MDS applied to Ardeidae baraminic distances and the stress of $k$-dimensional MDS on the same baraminic distance matrix plotted as a function of the number of dimensions ($k$). Ardeids are shown in black, outgroup taxa in gray.

## 2.17. Falconidae (Vertebrata: Aves: Falconiformes)

Dataset published by Griffiths (1999)

| | |
|---|---|
| Characters in published dataset: | 23 |
| Taxa in published dataset: | 25 |
| Character relevance cutoff: | 0.95 |
| Characters used to calculate BD: | 23 |
| Taxa used in BDC and MDS analysis: | 25 |
| Stress for 3D MDS: | 0.148 |
| $k_{min}$: | 6 |
| Median bootstrap value: | 66 |
| $F_{90}$: | 0.197 |

The Falconidae are one of the three families of the Falconiformes (the others are Accipitridae and Cathartidae), consisting of more than 60 species in approximately ten genera. Although falconids are generally well-known for their hunting abilities, the members of subfamily Polyborinae are scavengers. The true falcons of subfamily Falconinae are the keen hunters. Griffiths's (1999) analysis of falconid phylogeny supported the monophyly of the entire family and the two subfamilies. Two genera, *Micrastur* and *Herpetotheres*, were basal to the two falconid subfamilies. Griffiths's morphological dataset contained 23 syringeal characters for 25 taxa. The taxa consist of 20 falconids and three outgroups: the sharp-shinned hawk *Accipiter striatus*, the white pelican *Pelecanus onocrotalus*, and the eastern screech owl *Otus asio*.

All character states and taxa were used to calculate baraminic distances. The clearest group in the BDC results (Figure 36) is the Polyborinae. All polyborine taxon pairs have significant, positive BDC. Significant, negative BDC can be observed between polyborine and other falconids. The Falconinae are less well-defined. Most falconine taxa are positively correlated with each other, with the exception of *Polihierax insignis*, which is positively correlated with only six of the remaining twelve falconine taxa. The "basal" falconids *Micrastur* and *Herpetotheres* are positively correlated with ingroup and outgroup taxa. Both *Micrastur* species are positively correlated with *Accipiter*. *Herpetotheres* correlates positively with five taxa: *Polihierax semiorquatus*, *Microhierax erythrogonys*, *Spiziapteryx circumcinctus*, *Micrastur semiorquatus*, and *Accipiter striatus*. Bootstrap values are fairly poor (median 66%), and bootstrap values for all negative BDC are <90%.

66                    Animal and Plant Baramins

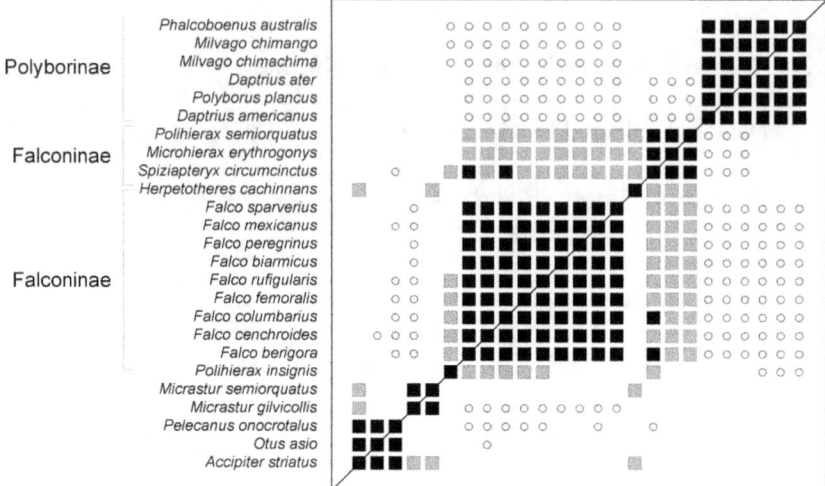

Figure 36. BDC bootstrap results for Falconidae, as calculated by BDISTMDS (relevance cutoff 0.95). Closed square indicate significant, positive BDC; open circles indicate significant, negative BDC. Black symbols indicate bootstrap values >90% in a sample of 100 pseudoreplicates. Grey symbols represent bootstrap values ≤90%.

The MDS results show a cluster of polyborines separated from a diffuse cluster of the remaining falconids and outgroup taxa (Figure 37). *Herpetotheres* and *Micrastur* are intermediate in position between the falconines and the outgroups. The two *Micrastur* species are closest to the outgroup taxa.

Given the extremely restricted nature of the characters (syringeal only), it is unlikely that these results reveal much about the baraminology of the falconids. Instead, the MDS and BDC results indicate that the syringes of most falconids are similar to other birds of prey and pelicans. The syringes of polyborines seem to be somewhat different from other falconids. Other than showing this similarity of syringes, these results seem to have few implications for falconid baraminology.

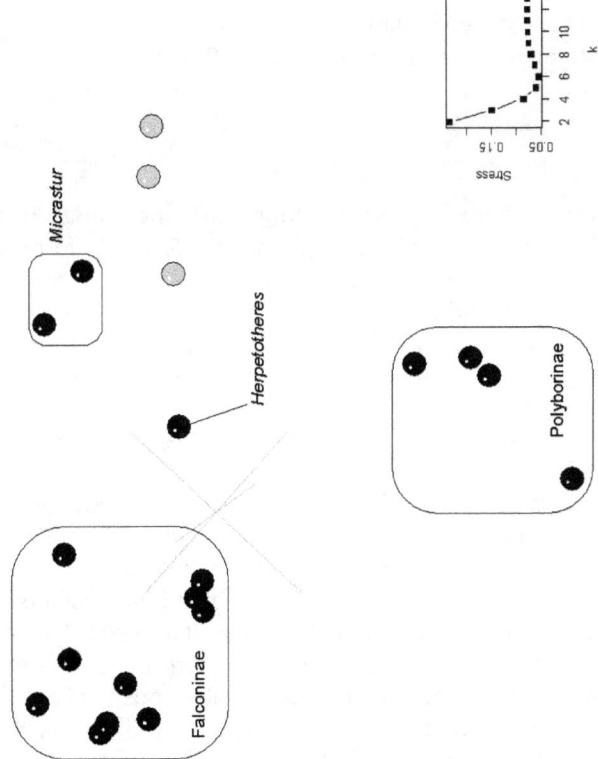

Figure 37. Three-dimensional MDS applied to Falconidae baraminic distances and the stress of *k*-dimensional MDS on the same baraminic distance matrix plotted as a function of the number of dimensions (*k*). Falconids are shown in black and outgroup taxa in gray.

## 2.18. Alcidae (Vertebrata: Aves: Charadriiformes)

Dataset published by Strauch (1985)

| | |
|---|---|
| Characters in published dataset: | 33 |
| Taxa in published dataset: | 23 |
| Character relevance cutoff: | 0.95 |
| Characters used to calculate BD: | 33 |
| Taxa used in BDC and MDS analysis: | 23 |
| Stress for 3D MDS: | 0.062 |
| $k_{min}$: | 4 |
| Median bootstrap value: | 72 |
| $F_{90}$: | 0.225 |

The Alcidae are a group of marine diving birds including auks, guillemots, and puffins, classified in the shorebird order Charadriiformes. Fourteen genera of alcids are placed into five tribes based on Strauch's (1985) phylogeny: Fraterculini (*Lunda, Fratercula, Cerorhinca*), Alcini (*Pinguinus, Uria, Alca, Alle*), Cepphini (*Synthliboramphus, Endomychura, Cepphus*), Brachyramphini (*Brachyramphus*), and Aethiini (*Ptychoramphus, Cyclorrhynchus, Aethia*). According to Baker et al.'s (2007) molecular phylogeny, the nearest outgroup is Stercorariidae, the skuas. Strauch's (1985) phylogeny was based on 33 characters and 23 taxa. The characters were primarily skeletal (21) but also included eight plumage/integumentary and four natural history characters. Strauch's 23 taxa are exclusively alcid. No outgroup taxa were included.

All taxa and characters were used to calculate baraminic distances. The tribe Fraterculini is the most distinguishable in the BDC results (Figure 38). The remaining tribes form a single group and are poorly differentiated. For example, the two Brachyramphini species correlate positively with all members of the Cepphini and some of the Aethiini. Likewise, the little auk *Alle alle* is positively correlated with all members of the Alcini, Brachyramphini, and the Cepphini. Significant, negative BDC occurs primarily between the Alcini and Aethiini, between the Fraterculini and Alcini, and between the Fraterculini and the Brachyramphini. No taxa in the Fraterculini are positively correlated with any other alcid taxa. Bootstrap values are somewhat poor, with highest bootstrap values (>90%) generally between members of the same tribe.

MDS results are quite good. The 3D stress is only 0.06, and the minimal stress of 0.04 is found at only four dimensions (Figure 39). The

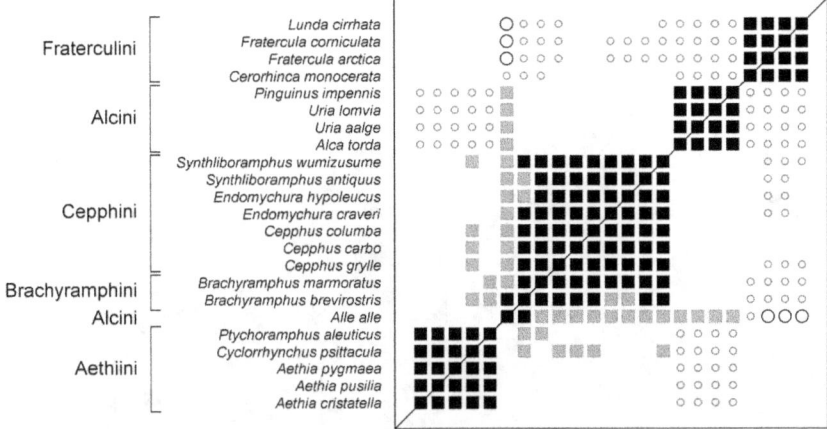

Figure 38. BDC bootstrap results for Alcidae, as calculated by BDISTMDS (relevance cutoff 0.95). Closed square indicate significant, positive BDC; open circles indicate significant, negative BDC. Black symbols indicate bootstrap values >90% in a sample of 100 pseudoreplicates. Grey symbols represent bootstrap values ≤90%.

distribution of alcid taxa in the MDS results is diffuse, with Fraterculini on one end, Alcini on the other, and the other three tribes in between.

The negative BDC between the Fraterculini and the remaining Alcidae could be weak evidence of a discontinuity, making the Alcidae (- Fraterculini) a holobaramin. This is not completely compatible with the molecular results of Baker et al. (2007), which places the Fraterculini in a clade with Aethiini. Given the lack of non-alcid outgroups in Strauch's dataset, it is impossible to determine if the negative BDC between Fraterculini and the remaining alcids is an indicator of discontinuity or simply an artifact of the small sample size. Nevertheless, given the present data, the Alcidae (- Fraterculini) appear to be a holobaramin.

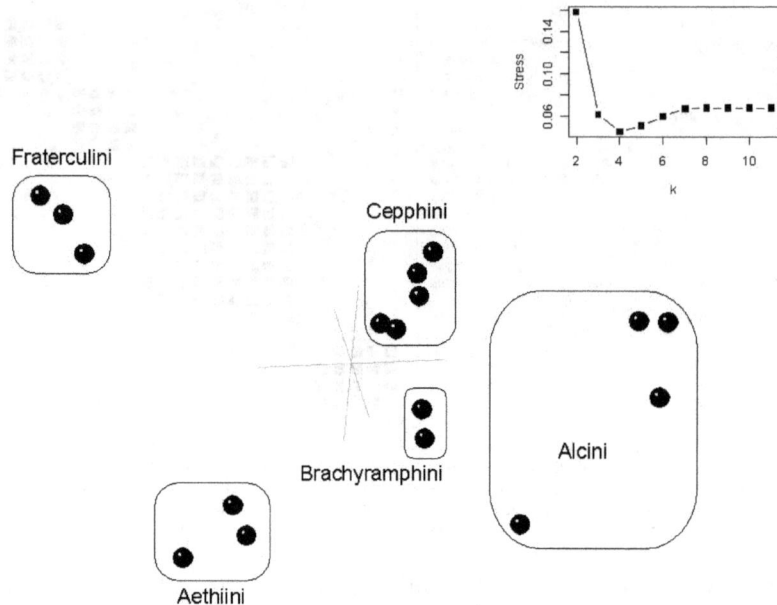

Figure 39. Three-dimensional MDS applied to Alcidae baraminic distances and the stress of $k$-dimensional MDS on the same baraminic distance matrix plotted as a function of the number of dimensions ($k$).

## 2.19. Fringillidae (Vertebrata: Aves: Passeriformes)

Dataset published by Chu (2002) and James (2004)

|  | Carduelinae (Chu 2002) | Drepanidinae (James 2004) |
|---|---|---|
| Characters in published dataset: | 225 | 84 |
| Taxa in published dataset: | 37 | 89 |
| Character relevance cutoff: | 0.95 | 0.9 |
| Characters used to calculate BD: | 177 | 45 |
| Taxa used in BDC and MDS analysis: | 37 | 89 |
| Stress for 3D MDS: | 0.455 | 0.243 |
| $k_{min}$: | 34 | 7 |
| Median bootstrap value: | 95 | 80 |
| $F_{90}$: | 0.632 | 0.371 |

The fringillid finches are part of a larger, putatively monophyletic group of passerine birds called "nine-primaried oscines" (e.g. see Klicka et al. 2000; Yuri and Mindell 2002). The family has been divided into at least five subfamilies: Carduelinae (finches), Drepanidinae (honeycreepers), Emberizinae (sparrows), Fringillinae (chaffinches), and Peucedraminae (the olive warbler). Even these subfamilies are not without controversy, as some authors treat various subfamilies or tribes as separate families (e.g. Ericson et al. 2000). Based on extensive hybridization, Fehrer (1993) classified the Carduelinae as a basic type (monobaramin) and suggested that the Emberizinae might also be a basic type. Wood's (2005a, pp. 117-118) analysis of the Galápagos finches included a brief and inconclusive discussion of fringillid baraminology.

Here, two datasets representing subsets of the Fringillidae are analyzed by BDC and MDS. The first focuses on the Carduelinae (Chu 2002) and the second on the Drepanidinae (James 2004). Chu's cardueline taxon sample consisted of 18 carduelines, 4 drepanidines, 14 other fringillids, and one outgroup (Estrildidae: *Lagonosticta*), for a total of 37 taxa. The 225 characters of Chu's dataset were craniomandibular (105), postcranial (43), and plumage/integumentary (77). James's drepanidine dataset sampled 89 taxa: 35 extant drepanidines, 24 fossil and subfossil drepanidines, 29 other fringillids (including 18 carduelines), and one outgroup (Passeridae: *Passer*). James's 84 characters included 72 craniomandibular, 7 hyoid, and 5 postcranial characters. For both datasets, all taxa were retained for baraminic distance calculations.

For the Chu dataset, a character relevance cutoff of 0.95 retained 177 characters for baraminic distance calculation (95 craniomandibular, 37 postcranial, and 43 plumage/integumentary). One large block of taxa consisting of the carduelines, the two *Fringilla* species, and the drepanidine *Telespiza cantans* is evident from the BDC results (Figure 40). Significant, positive BDC is extensive within the Carduelinae + *Telespiza* and more sporadic between the *Fringilla* species and the carduelines. Most of the BDC within the Carduelinae + *Telespiza* have bootstrap values >90%, while the positive BDC connecting *Fringilla* to this group has bootstrap values <90%. The two species of carduelines in the *Coccothraustes* genus are excluded from the Carduelinae + *Telespiza* group. The other three drepanidine species also form a group with significant, positive BDC, as do the remaining fringillid species. The estrildid outgroup *Lagonosticta* correlates negatively with the two *Coccothraustes* species but neither positively nor negatively with any other taxa. Significant, negative BDC is sparse, most frequently occuring between the *Coccothraustes* species and the non-cardueline fringillids. Bootstrap values are generally quite high for this dataset (median 95%).

The MDS for Chu's cardueline dataset is exceptionally poor. The 3D stress is 0.455, and the minimum stress of 0.001 occurs at 34 dimensions. Consequently, the 3D results are likely a poor representation of the true distribution of taxa. Nevertheless, taxon distribution patterns consistent with the BDC results can be detected (Figure 41). The Carduelinae + *Telespiza* form a tight cluster, with *Coccothraustes* separated to the side. The outgroup *Lagonosticta* is closest to the non-*Telespiza* drepanidines and somewhat centrally located between the Carduelinae + *Telespiza* and the other fringillids.

With a character relevance cutoff of 0.9, 45 craniomandibular characters were used to calculate baraminic distances from James's Drepanidinae dataset. There seem to be two large groups of taxa in the BDC results (Figure 42), but they do not correspond to any recognized classification and they heavily overlap. A few taxa, such as *Psittirostra psittacea* and *Melamprosops phaeosoma* are correlated with members of both groups. Though significant, negative BDC occurs between the two groups, most of it has bootstrap values <90%. The positive BDC within each group generally has bootstrap values >90%. The outgroup *Passer domesticus* is positively correlated with 46 of the 88 ingroup taxa and negatively correlated with 37 ingroup taxa. The BDC results do not separate the drepanidines from the other fringillids.

As seen in the cardueline dataset, the MDS results for the drepanidines are somewhat poor. The 3D stress is 0.243, but the minimum stress of 0.156 occurs at only 7 dimensions, indicating that the three-dimensional results are not that far from the optimal distribution of taxa. The fringillids and the outgroup *Passer* form a single group with three notable outliers in the MDS results (Figure 43). The outliers are fossil taxa: the Maha'ulepu finch, the Maui *Chloridops* species, and the Hawaii *Hemignathus* sp. The drepanidines and other fringillids are thoroughly mixed, and there is little evidence of the two groups observed in the BDC results.

The widespread positive BDC observed for both datasets implies that all fringillid taxa examined were members of a single holobaramin. Neither BDC nor MDS could detect a significant difference between the subfamilies of Fringillidae. If Fringillidae itself is a holobaramin, then inclusion of non-fringillid outgroups should have detected a discontinuity.

Given the presence of nonfringillid outgroups in both datasets (*Lagonosticta* for the carduelines and *Passer* for the drepanidines), the failure to detect this potential discontinuity is notable. This could be due to a true lack of discontinuity (suggesting that the Passeridae, Estrildidae, and Fringillidae are all members of a larger holobaramin) or to spurious character sampling. It is interesting to note that the cardueline dataset does not show any positive BDC between the outgroup and ingroup taxa, and the sampled characters include craniomandibular, postcranial and plumage/integumentary characters. The drepanidine dataset omits the non-cranial characters due to low relevance, and the outgroup shows extensive positive correlation with fringillid taxa. Since baraminological studies should be based on holistic data, it is interesting to see a less holistic dataset showing apparent "continuity" between ingroup and outgroup taxa.

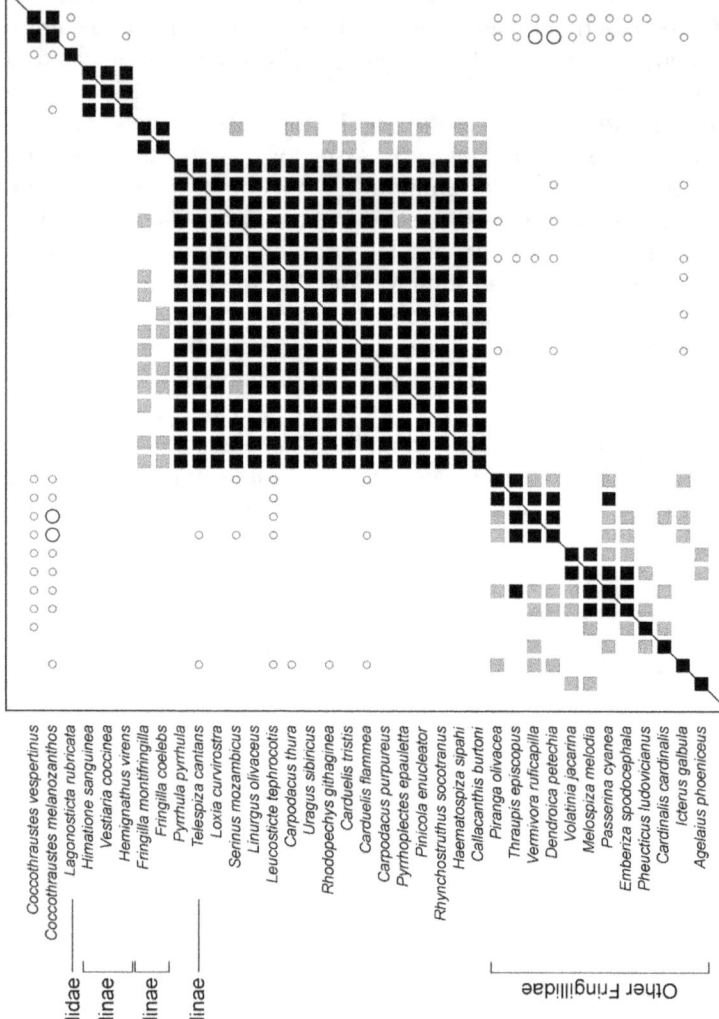

Figure 40. BDC bootstrap results for Carduelinae, as calculated by BDISTMDS (relevance cutoff 0.95). Closed square indicate significant, positive BDC; open circles indicate significant, negative BDC. Black symbols indicate bootstrap values >90% in a sample of 100 pseudoreplicates. Grey symbols represent bootstrap values ≤90%.

Animals 75

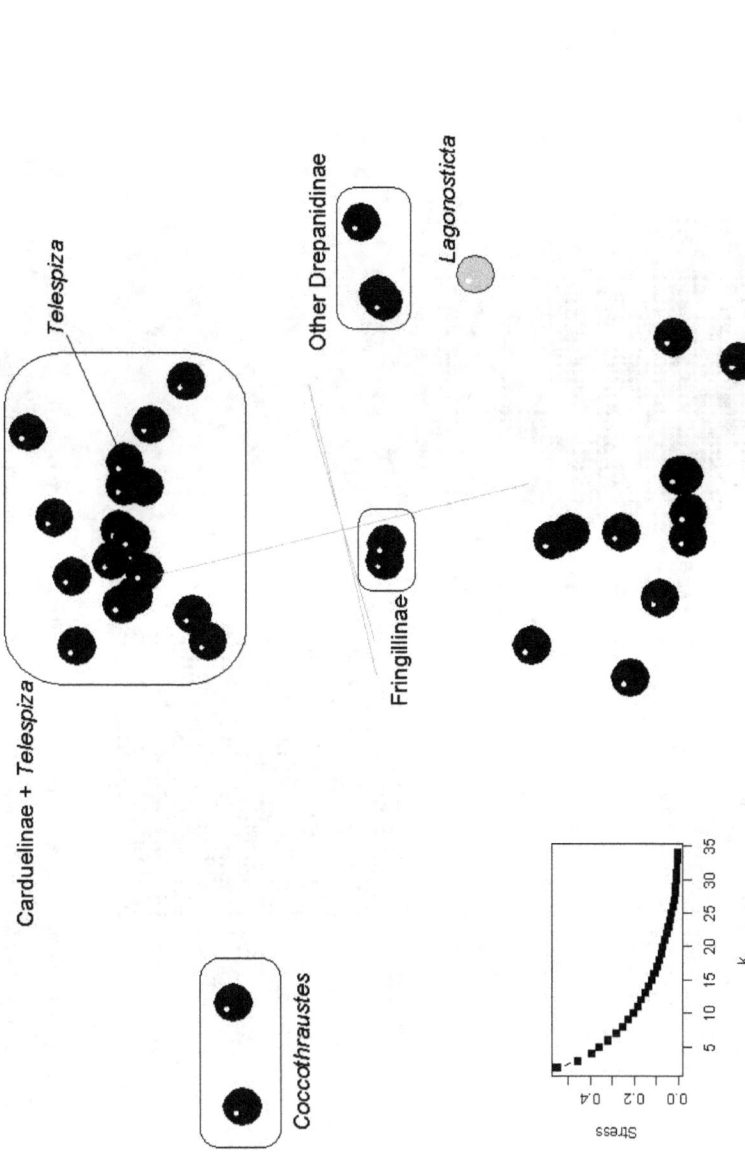

Figure 41. Three-dimensional MDS applied to Carduelinae baraminic distances and the stress of $k$-dimensional MDS on the same baraminic distance matrix plotted as a function of the number of dimensions ($k$). Carduelines are shown in black and outgroup taxa in gray.

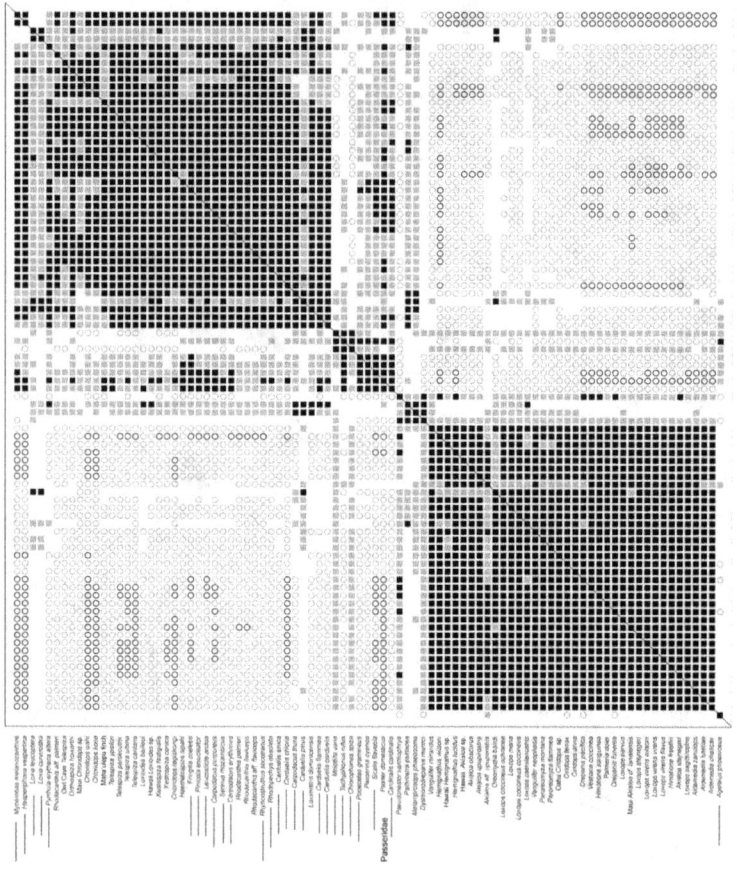

Figure 42. BDC bootstrap results for Drepanidinae, as calculated by BDISTMDS (relevance cutoff 0.9). Closed square indicate significant, positive BDC; open circles indicate significant, negative BDC. Black symbols indicate bootstrap values > 90% in a sample of 100 pseudoreplicates. Grey symbols represent bootstrap values ≤90%. Non-drepanidine fringillids are indicated by horizontal lines.

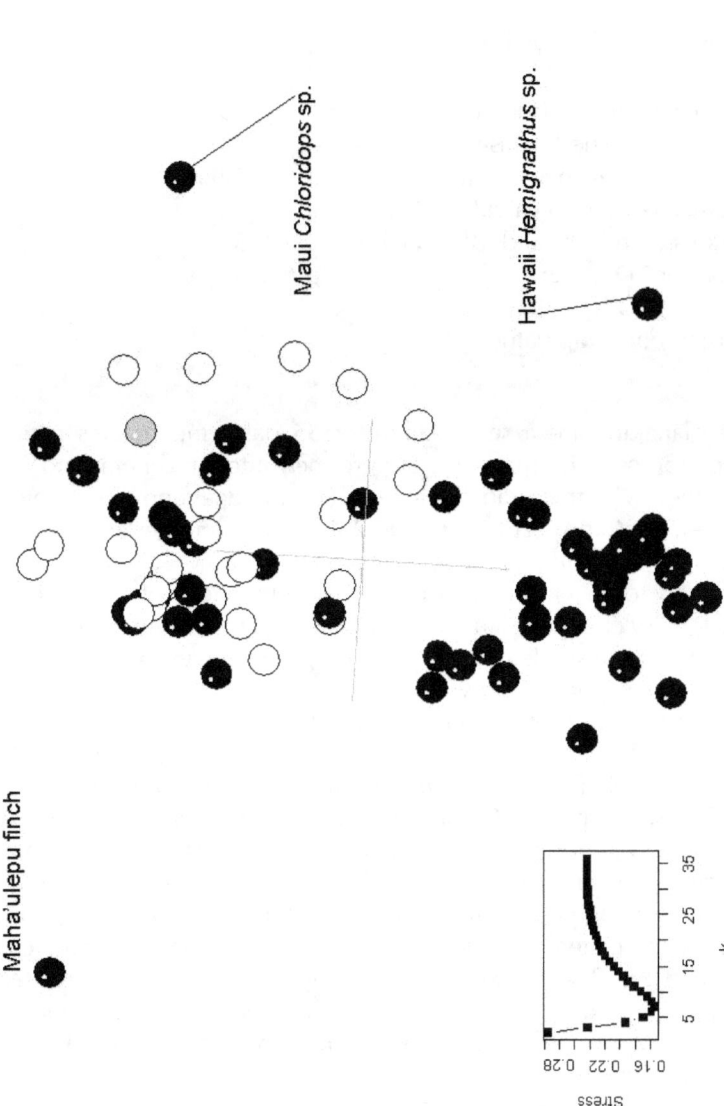

Figure 43. Three-dimensional MDS applied to Drepanidinae baraminic distances and the stress of $k$-dimensional MDS on the same baraminic distance matrix plotted as a function of the number of dimensions ($k$). Drepanidines are shown in black, other fringillids in white, and the outgroup taxon in gray.

## 2.20. Pipridae (Vertebrata: Aves: Passeriformes)

Dataset published by Prum (1992)

| | |
|---|---|
| Characters in published dataset: | 59 |
| Taxa in published dataset: | 38 |
| Character relevance cutoff: | 0.95 |
| Characters used to calculate BD: | 58 |
| Taxa used in BDC and MDS analysis: | 38 |
| Stress for 3D MDS: | 0.139 |
| $k_{min}$: | 8 |
| Median bootstrap value: | 81 |
| $F_{90}$: | 0.361 |

The manakins are a small family of 55 passerine species native to the Neotropics. The manakins are grouped into 11 genera, and the family is allied with the tyrant flycatchers (Tyrannidae) and the cotingas (Cotingidae). McCarthy (2006, pp. 204-205) lists six intergeneric hybrids for this family involving six genera, indicating that at least part of the family is likely to be a monobaramin. Prum's phylogeny of the Pipridae was based on 59 characters and 38 taxa. The taxa represented all eleven piprid genera plus a single outgroup taxon. The characters were entirely based on syringeal morphology.

All taxa and 58 of the 59 characters were used to calculate baraminic distances. The BDC results show extensive positive BDC (Figure 44). As in Prum's (1992) phylogeny, the *Pipra* species were separated into distinct groups. One set of species, *P. cornuta, P. erythrocephala, P. mentalis, P. rubrocapilla, P. chloromeros, P. filicarda, P. fasciicauda,* and *P. aureola,* is separated from the remaining piprid species, including seven other *Pipra* species by significant, negative BDC. One final *Pipra, P. pipra,* is positively correlated with both groups of taxa (as are *Heterocercus linteatus* and *H. flavivertex*). The outgroup is positively correlated with 27 piprids and negatively correlated with seven. Bootstrapping is moderately good (median 81%) and especially good (>90%) for the significant, positive BDC.

The MDS results are also quite good (3D stress 0.139) and reveal a diffuse, roughly triangular distribution of taxa (Figure 45). The outgroup taxon is found near the center of the piprids. Though the outlying *Pipra* species (*P. cornuta* et al.) are slightly separated from the main group of Pipridae, they appear to be merely an extension of the main cluster rather than a separate cluster of their own.

As in the case of the Falconidae, the wholly syringeal characters are likely to be unsuitable for baraminological research. The most reasonable conclusion we can reach from these results is that the outlying *Pipra* species have a differently shaped syrinx than the other piprids. Likewise, the piprid syrinx does not seem to differ significantly from the outgroup syrinx. Given the hybridization and the similarity of the syrinx, we might tentatively conclude that the Pipridae are minimally a monobaramin.

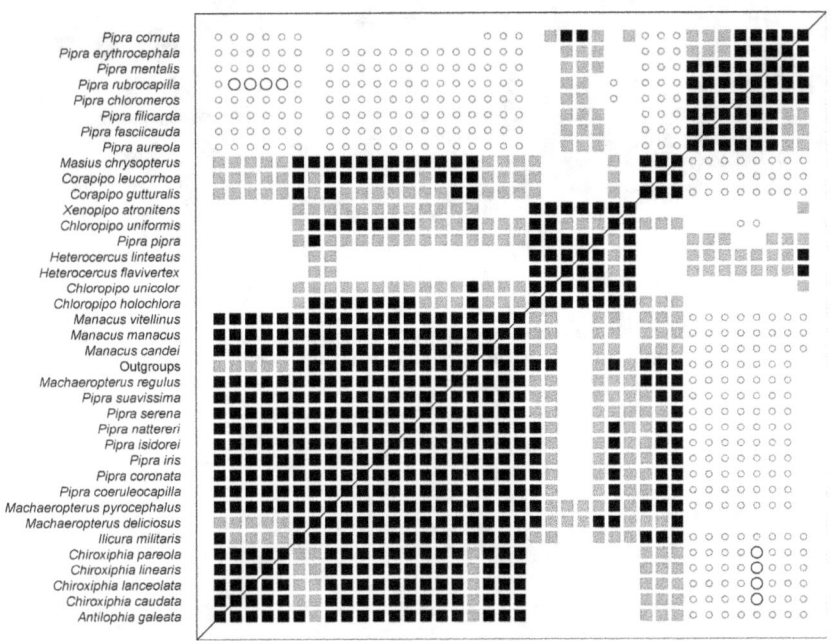

Figure 44. BDC bootstrap results for Pipridae, as calculated by BDISTMDS (relevance cutoff 0.95). Closed square indicate significant, positive BDC; open circles indicate significant, negative BDC. Black symbols indicate bootstrap values >90% in a sample of 100 pseudoreplicates. Grey symbols represent bootstrap values ≤90%.

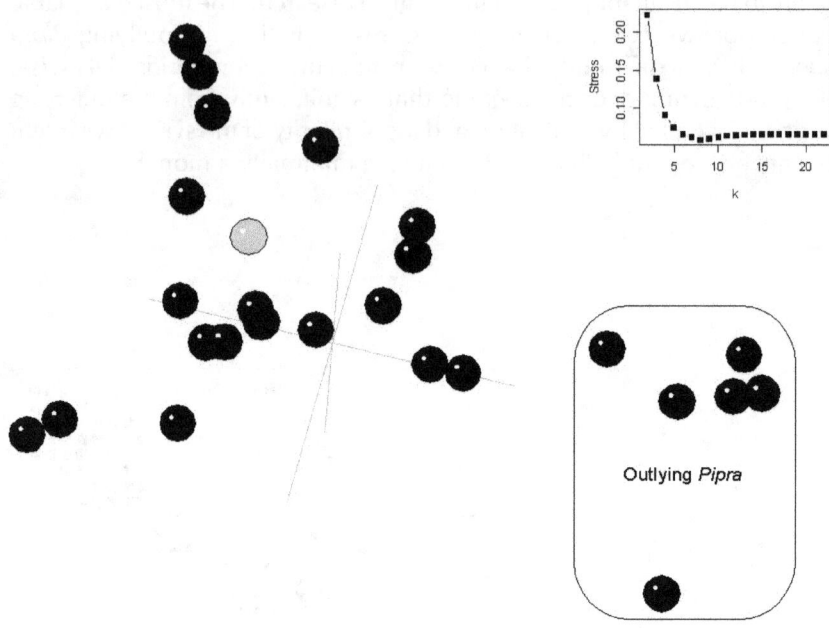

Figure 45. Three-dimensional MDS applied to Pipridae baraminic distances and the stress of *k*-dimensional MDS on the same baraminic distance matrix plotted as a function of the number of dimensions (*k*). Piprids are shown in black, the outgroup taxon in gray.

## 2.21. Spheniscidae (Vertebrata: Aves: Sphenisciformes)

Dataset published by Ksepka et al. (2006)

| | |
|---|---|
| Characters in published dataset: | 181 |
| Taxa in published dataset: | 50 |
| Character relevance cutoff: | 0.95 |
| Characters used to calculate BD: | 109 |
| Taxa used in BDC and MDS analysis: | 29 |
| Stress for 3D MDS: | 0.089 |
| $k_{min}$: | 7 |
| Median bootstrap value: | 100 |
| $F_{90}$: | 0.855 |

The penguins are extremely recognizable water birds, thanks in part to their ubiquity in American popular culture. Eighteen species of extant penguins are grouped into six genera of the family Spheniscidae and the order Sphenisciformes. Wood's (2005a, pp. 125-133) previous analysis of penguin baraminology was based on a set of 33 plumage/integumentary and natural history characters. The penguins appeared to be a monobaramin, with the exception of the king and emperor penguins (genus *Aptenodytes*), for which evidence of continuity with other penguins was lacking. Evidence of discontinuity with nonpenguins was good, with significant, negative BDC observed between the twelve outgroup taxa and most of the penguins. Wood tentatively identified the penguins as a holobaramin.

Ksepka et al.'s (2006) phylogenetic study of the Sphenisciformes was based in part on analysis of a morphological dataset consisting of 74 plumage/integumentary and natural history characters, 35 craniomandibular, 56 postcranial, and 16 soft tissue characters. Ksepka et al. extensively sampled the taxa of Sphenisciformes, including all the extant penguins as well as 21 fossil sphenisciforms and eleven outgroups (ten procellariiforms and the loon *Gavia immer*), for a total of fifty taxa.

Due to extremely low taxic relevance (0.05-0.436, median 0.122), all of the fossil sphenisciforms from Ksepka et al.'s dataset were omitted from the baraminic distance calculations. With the remaining 29 taxa at a character relevance cutoff of 0.95, 109 of the 181 characters were used for baraminic distance calculations. These characters consisted of 73 plumage/integumentary and natural history characters, 33 craniomandibular, and three postcranial characters. Despite these lost characters, the BDC results are quite clear (Figure 46). All penguin taxa exhibit significant, positive BDC with each other and significant, negative BDC with the outgroups. Bootstrap values were extremely high (median 100%). The MDS results were also quite good (3D stress 0.089). The penguins form a somewhat linear structure separated from the outgroup taxa (Figure 47).

Given the fairly holistic character set and high degree of morphological similarity apparent even to non-experts, there seems little remaining doubt that the penguins form a single holobaramin. This dataset had a broader sampling of characters than the one used by Wood (2005a), and consequently, continuity between *Aptenodytes* and the remaining penguins could be established by significant, positive BDC. While it would be good to eventually include the fossil sphenisciforms omitted from this study in a future baraminological analysis, the present results seem unambiguous: All extant penguins belong to a single created kind.

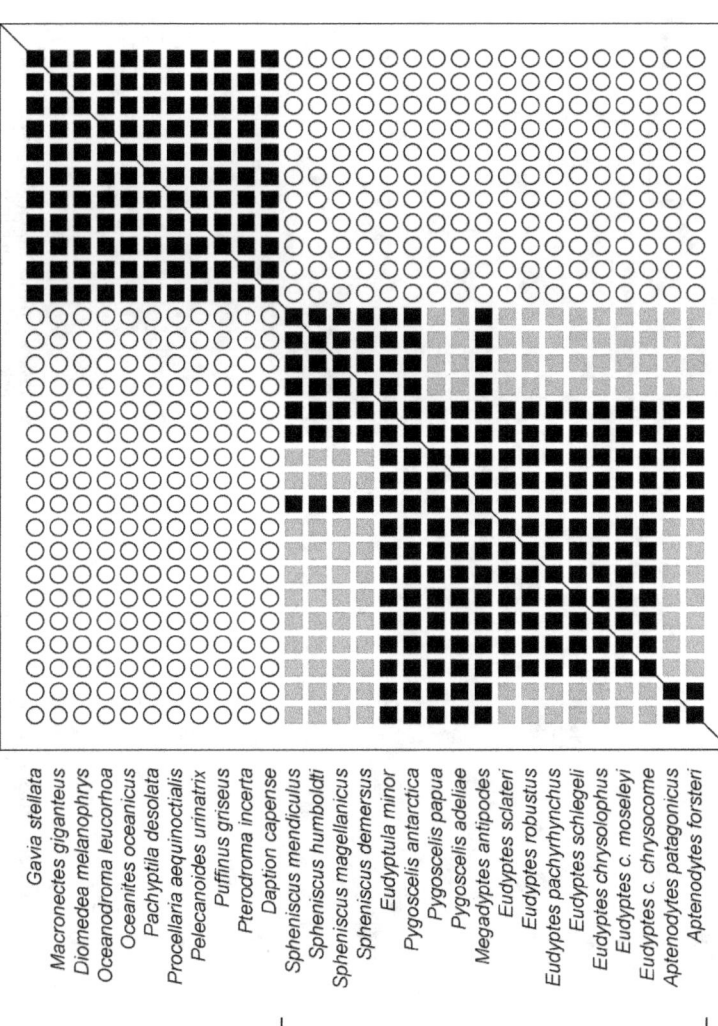

Figure 46. BDC bootstrap results for Spheniscidae, as calculated by BDISTMDS (relevance cutoff 0.95). Closed square indicate significant, positive BDC; open circles indicate significant, negative BDC. Black symbols indicate bootstrap values > 90% in a sample of 100 pseudoreplicates. Grey symbols represent bootstrap values ≤90%.

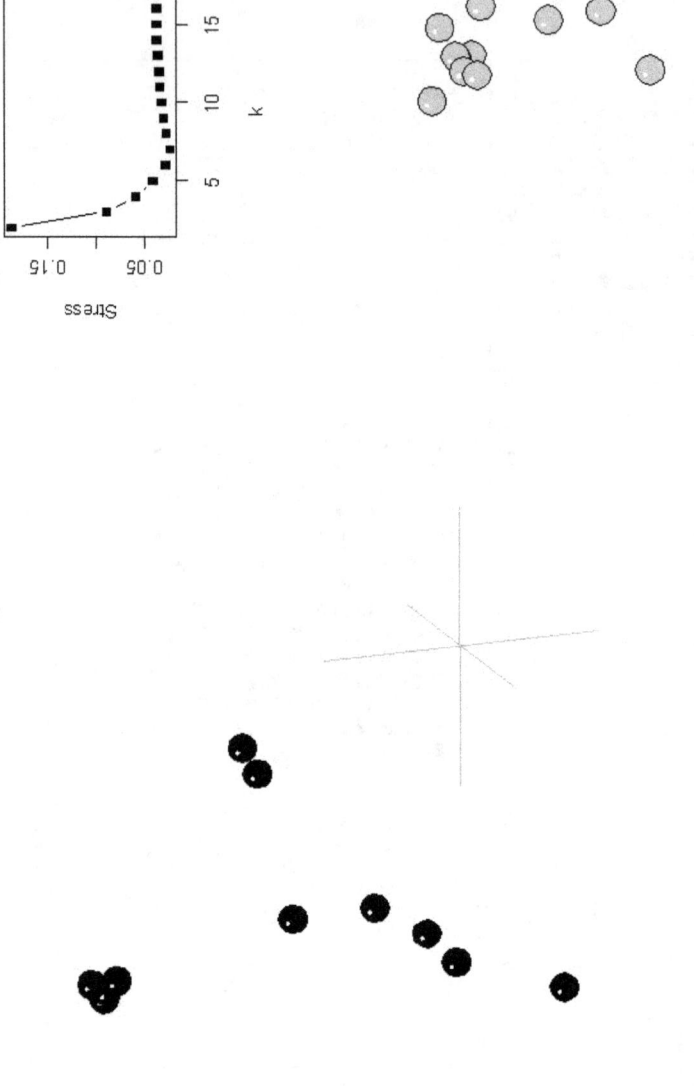

Figure 47. Three-dimensional MDS applied to Spheniscidae baraminic distances and the stress of *k*-dimensional MDS on the same baraminic distance matrix plotted as a function of the number of dimensions (*k*). Spheniscids are shown in black, outgroup taxa in gray.

## 2.22. Pygopodidae (Vertebrata: Reptilia: Squamata)

Dataset published by Kluge (1976)

| | |
|---|---|
| Characters in published dataset: | 86 |
| Taxa in published dataset: | 22 |
| Character relevance cutoff: | 0.95 |
| Characters used to calculate BD: | 86 |
| Taxa used in BDC and MDS analysis: | 22 |
| Stress for 3D MDS: | 0.106 |
| $k_{min}$: | 7 |
| Median bootstrap value: | 100 |
| $F_{90}$: | 0.732 |

Family Pygopodidae consists of 38 species of legless lizards of Australia and New Guinea (Jennings et al. 2003). Forelimbs are absent in all species, but hindlimbs vary from small paddle-like limbs to completely absent. Pygopodids are classified into six genera and two subfamilies, Pygopodinae (*Pygopus*, *Delma*) and Lialisinae (*Lialis*, *Pletholax*, *Ophidiocephalus*, *Aprasia*). Kluge's (1976) analysis of Pygopodidae phylogeny was based on 86 characters and 22 taxa. The characters were 60 craniomandibular and 26 postcranial. The taxa were 21 of the 38 pygopodid species, including representatives of all genera, and the hypothetical ancestral pygopodid comprised of the estimated "primitive" states of all characters.

For baraminic distance calculations, all characters were used at the character relevance cutoff of 0.95. All taxa were used as well. There are two well-defined groups evident in the BDC results, one of which corresponds to the subfamily Pygopodinae plus the genus *Lialis* and the other to the subfamily Lialisinae without the *Lialis* species (Figure 48). Within each group, taxon pairs generally have significant, positive BDC, and between the two groups, taxon pairs have significant, negative BDC. Bootstrap values are extremely high (median 100%) for both positive and negative BDC.

The MDS results reveal a more diffuse distribution of taxa than indicated in the BDC results. The Pygopodinae and the hypothetical pygopodid ancestor form a tight cluster, while the remaining genera of Lialisinae do not form a single cluster (Figure 49). Each lialisine genus forms a separate cluster, with *Lialis* closest to the pygopodines.

Based on significant, positive BDC, the Pygopodinae + *Lialis* appear to be a monobaramin and possibly a holobaramin. The

86　　　　　　　　Animal and Plant Baramins

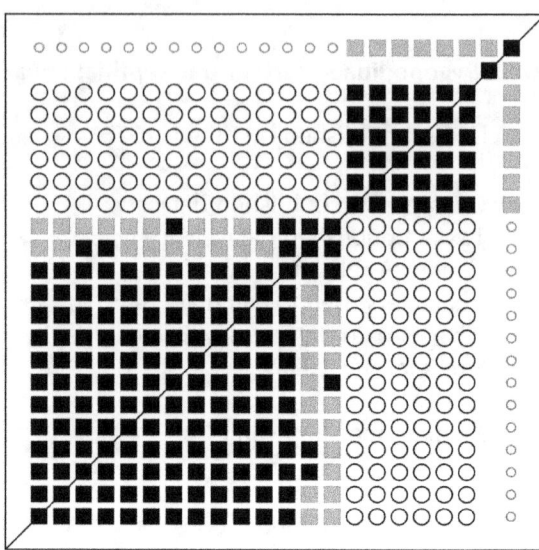

Figure 48. BDC bootstrap results for Pygopodidae, as calculated by BDISTMDS (relevance cutoff 0.95). Closed square indicate significant, positive BDC; open circles indicate significant, negative BDC. Black symbols indicate bootstrap values >90% in a sample of 100 pseudoreplicates. Grey symbols represent bootstrap values ≤90%.

separation between these taxa and the remaining Lialisinae could be a discontinuity, as evidenced by the significant, negative BDC. The MDS results support the grouping of all pygopodines in a single monobaramin, although *Lialis* species are separated from the Pygopodinae. Although contradicting the phylogenetic study of Kluge (1976), which showed a monophyletic Lialisinae and polyphyletic Pygopodinae, a group consisting of Pygopodinae + *Lialis* would be consistent with the molecular phylogenetic studies of Jennings et al. (2003), which showed that these groups were monophyletic. While it would be preferable to analyze a true non-pygopodid outgroup taxon in order to verify the intrafamilial discontinuity observed here, the present results support a holobaramin of Pygopodinae + *Lialis*.

# Animals

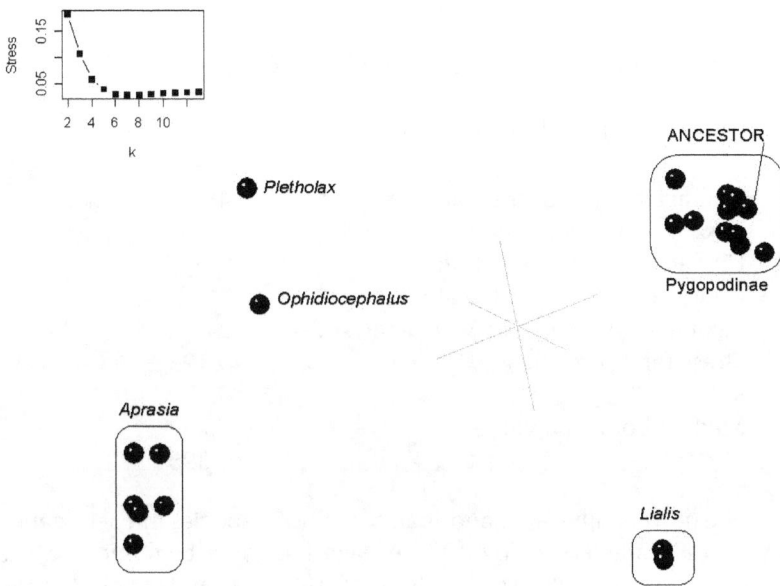

Figure 49. Three-dimensional MDS applied to Pygopodidae baraminic distances and the stress of $k$-dimensional MDS on the same baraminic distance matrix plotted as a function of the number of dimensions ($k$).

### 2.23. Salamandridae (Vertebrata: Amphibia: Caudata)

Dataset published by Titus and Larson (1995)

| | |
|---|---|
| Characters in published dataset: | 48 |
| Taxa in published dataset: | 21 |
| Character relevance cutoff: | 0.9 |
| Characters used to calculate BD: | 36 |
| Taxa used in BDC and MDS analysis: | 21 |
| Stress for 3D MDS: | 0.123 |
| $k_{min}$: | 5 |
| Median bootstrap value: | 76.5 |
| $F_{90}$: | 0.395 |

The newt family Salamandridae contains 53 species in 13-15 genera. Newts are characterized by a three-stage life cycle beginning with an aquatic larval stage, followed by a terrestrial eft stage, and concluding with an aquatic or semiaquatic adult. Salamandrids are found in the northern hemisphere of the Old and New Worlds. The most common North American species is the red-spotted newt (*Notophthalmus viridescens*) which is found in nearly any freshwater pond or lake in the eastern half of the United States and Canada (Petranka 1998, pp. 451-462). Titus and Larson's (1995) analysis of salamandrids was based in part on a set of 48 characters scored for 18 salamandrid species (representing 15 genera) and three outgroups. The characters came from 6 cranial, 13 postcranial, 19 soft tissue, and 10 reproductive attributes. Their morphological and molecular phylogenies supported a monophyletic Salamandridae.

All taxa and 36 characters (6 cranial, 11 postcranial, 10 soft tissue, and 9 reproductive) were used to calculate baraminic distances (0.9 character relevance cutoff). There are three groups evident in the BDC results (Figure 50). One group consists of the *Mertensiella*, *Salamandra*, and *Chioglossa* species, a second group contains *Tylototriton*, *Pleurodeles*, and the outgroups, and the remaining salamandrids form the final group. These groups are defined by significant, positive BDC for nearly all taxon pairs. Within the two salamandrid-only groups, bootstrap values are generally >90%. Some species correlate with more than one group. For example, the rough skinned newt *Taricha granulosa* positively correlates with all *Tylototriton* and *Pleurodeles* species, and the outgroup *Ambystoma gracile* is positively correlated with all species of *Salamandra* and *Mertensiella*. The Italian spectacled salamander *Salamandrina terdigitata* is not correlated positively or negatively with any other taxa. Significant, negative BDC is universal between members of the *Mertensiella/Salamandra/Chioglossa* group and the other salamandrid-only group. For comparisons involving *Mertensiella* or *Salamandra* and one of the other salamandrids, significant, negative BDC had bootstrap values >90%. Negative correlation with outgroup taxa was rare, occurring only between the northwestern salamander *Ambystoma gracile* and seven salamandrids. No significant, negative BDC with *A. gracile* had bootstrap values >90%. *A. gracile* also had significant, positive BDC with all species of *Tylototriton* as well as *Salamandra* and *Mertensiella*.

The three groups of taxa detected in the BDC results are not as apparent in the MDS results (Figure 51). The salamandrid species are distributed in a wide arc, with the outgroup taxa in the middle. While *Mertensiella* and *Salamandra* are clustered tightly and widely separated from the rest of the salamandrids, *Chioglossa* is offset more towards the rest of the salamandrids. *Salamandrina* lies between *Chioglossa* and the main salamandrid group. *Tylototriton* and *Pleurodeles* also cluster together and are adjacent to the three outgroup taxa.

These results seem to imply that some of the salamandrids share continuity with ambystomatids and dicamptodontids, based on significant, positive BDC and the distribution of taxa in MDS. The negative BDC might suggest a discontinuity between two groups of salamandrids, but significant, positive BDC connects them both to the third group of salamandrid and outgroup taxa. The dataset is also more holistic than many datasets in this study, even though the characters are fewer. A conservative conclusion would place the salamandrids in a single monobaramin with a tentative classification of the ambystomatids and dicamptodontids in the same monobaramin.

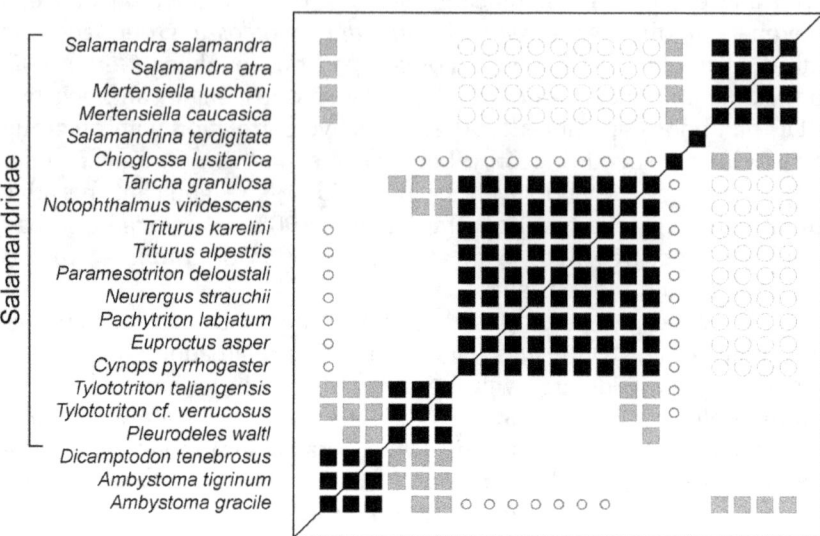

Figure 50. BDC bootstrap results for Salamandridae, as calculated by BDISTMDS (relevance cutoff 0.9). Closed square indicate significant, positive BDC; open circles indicate significant, negative BDC. Black symbols indicate bootstrap values >90% in a sample of 100 pseudoreplicates. Grey symbols represent bootstrap values ≤90%.

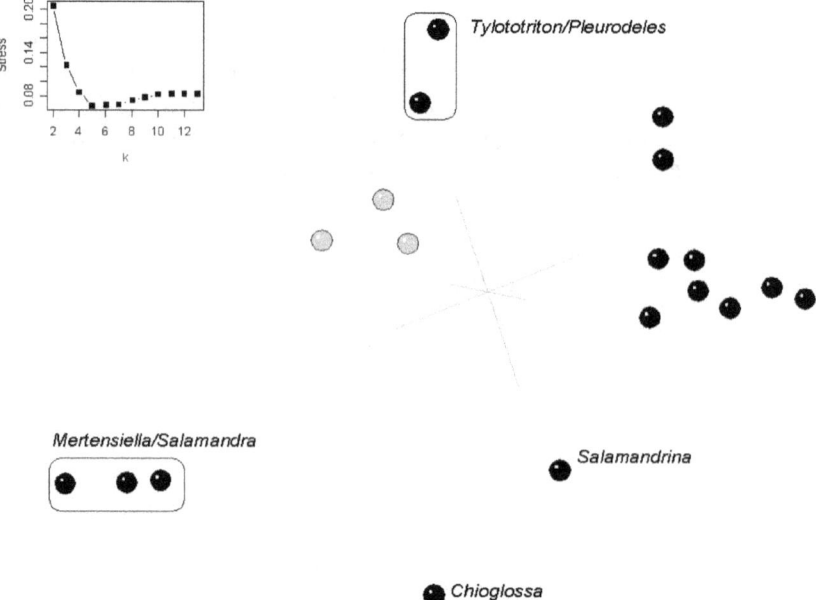

Figure 51. Three-dimensional MDS applied to Salamandridae baraminic distances and the stress of $k$-dimensional MDS on the same baraminic distance matrix plotted as a function of the number of dimensions ($k$). Salamandrids are shown in black and outgroup taxa in gray.

## 2.24. Gadidae (Vertebrata: Osteichthyes: Gadiformes)

Dataset published by Teletchea et al. (2006)

| | |
|---|---|
| Characters in published dataset: | 30 |
| Taxa in published dataset: | 23 |
| Character relevance cutoff: | 0.95 |
| Characters used to calculate BD: | 23 |
| Taxa used in BDC and MDS analysis: | 23 |
| Stress for 3D MDS: | 0.104 |
| $k_{min}$: | 4 |
| Median bootstrap value: | 93 |
| $F_{90}$: | 0.534 |

The family Gadidae is comprised of cod, haddock, and pollock, which occupy coastal zones of the northern oceans. Various experts have placed 22, 21, 15, and 12 genera in the Gadidae. Teletchea et al. (2006) placed 21 genera in the family Gadidae based on molecular and morphological phylogenetic analyses. Their morphological dataset consisted of 19 fin and skeletal, 6 soft tissue, and 5 life history characters. They sampled all 22 genera of fish previously referred to Gadidae, with the codling family Moridae as the outgroup. From their phylogenetic results, they recognized four subfamilies: Gadinae (12 genera), Phycinae (*Phycis*, *Urophycis*, and *Raniceps*), Gaidropsarinae (*Gaidropsarus*, *Ciliata*, and *Enchelyopus*), and Lotinae (*Lota*, *Molva*, and *Brosme*).

All taxa and 23 characters (13 fins/skeletal, 6 soft tissue, 4 life history) were retained for baraminic distance calculations. There are two very clear groups in the BDC results (Figure 52). The first group corresponds to Teletchea et al.'s (2006) Gadinae. Within this group, all taxa share significant, positive BDC with bootstrap values >90%. The second group, consisting of the remaining taxa, has fewer instances of significant, positive BDC between taxa and the bootstrap values are generally <90%. Comparisons of gadines with non-gadines reveals either significant, negative BDC or nonsignificant BDC. No significant, positive BDC connects a gadine to any non-gadine species.

The MDS results corroborate the two groups detected in the BDC analysis (Figure 53). The gadine taxa form a tight cluster well separated from the remaining taxa. The other non-gadine taxa have a more diffuse distribution, with the outgroup Moridae somewhat centrally located in that group.

Given the holistic nature of the character set, there seems little room to question the holobaraminic status of the Gadinae. Continuity between gadine species is confirmed by the significant, positive BDC and by the tight clustering in MDS. Discontinuity between gadines and non-gadines is supported by significant, negative BDC and the separation observed in the MDS results. As mentioned above, Teletchea et al. (2006) placed twelve genera in the Gadinae, which happens to correspond to the circumscription of the entire gadid family by Howes (1991). Going with Howes's classification, then, the family Gadidae *sensu stricto* is a holobaramin.

Animals 93

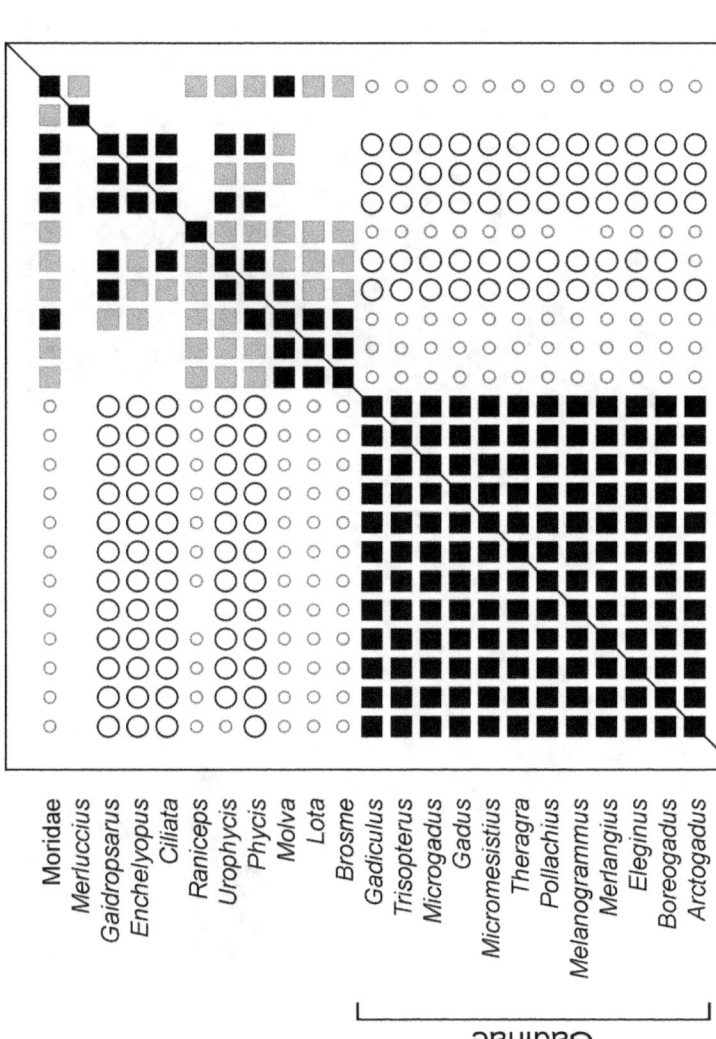

Figure 52. BDC bootstrap results for Gadidae, as calculated by BDISTMDS (relevance cutoff 0.95). Closed square indicate significant, positive BDC; open circles indicate significant, negative BDC. Black symbols indicate bootstrap values >90% in a sample of 100 pseudoreplicates. Grey symbols represent bootstrap values ≤90%.

94 Animal and Plant Baramins

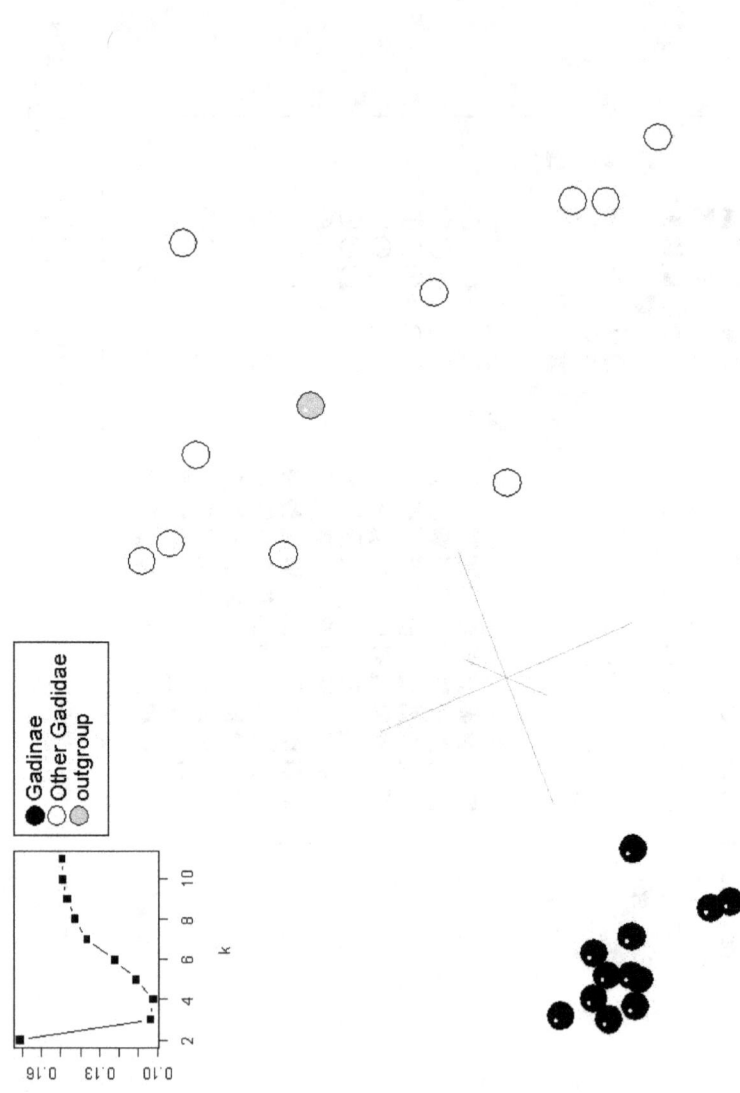

Figure 53. Three-dimensional MDS applied to Gadidae baraminic distances and the stress of $k$-dimensional MDS on the same baraminic distance matrix plotted as a function of the number of dimensions ($k$). Members of subfamily Gadinae are shown in black, other gadids in white, and the outgroup in gray.

## 2.25. Liparidae (Vertebrata: Osteichthyes: Scorpaeniformes)

Dataset published by Knudsen et al. (2007)

| | |
|---|---|
| Characters in published dataset: | 84 |
| Taxa in published dataset: | 29 |
| Character relevance cutoff: | 0.95 |
| Characters used to calculate BD: | 61 |
| Taxa used in BDC and MDS analysis: | 29 |
| Stress for 3D MDS: | 0.116 |
| $k_{min}$: | 5 |
| Median bootstrap value: | 93 |
| $F_{90}$: | 0.557 |

Liparidae is a large family consisting of more than 300 species of snailfish in 29 genera found in the North Pacific, North Atlantic, Antarctic, and Arctic oceans. Snailfish have a distinct morphology with large pectoral fins, no scales, and a mucous-covered skin. Liparids are thought to be closely allied to the lumpsuckers, family Cyclopteridae. Knudsen et al.'s (2007) molecular and morphological phylogenetic study supported the monophyly of the family and suggested that the deep-water species were derived from shallow-water forms. Their morphological data consisted of 84 body shape characters from 24 liparids and 5 cyclopterids. The 24 liparids represent only nine of the 29 genera.

All taxa and 61 of the 84 characters were used to calculate baraminic distances. Two clear groups are evident in the BDC results (Figure 54), the Liparidae and the Cyclopteridae. Within each group, taxon pairs generally have significant, positive BDC with high (>90%) bootstrap values. Between the families, significant, negative BDC occurs for 75 of the 120 taxon pairs. It should be noted that the Liparids could potentially be subdivided into two overlapping groups based on a lack of significant BDC. The first group consists of the *Liparis* species, which do not positively correlate with 14 of the remaining 18 liparid taxa. Two species of *Careproctus* correlate positively with members of both groups. One taxon, the variegated snail fish *Liparis gibbus* correlates positively with all members of the outgroup Cyclopteridae and negatively with eleven lipard taxa.

Clusters of taxa corresponding to the Liparidae and Cyclopteridae are also visible in the MDS results (Figure 55). The liparids have a somewhat diffuse distribution, and the cyclopterids are clustered separately from the lipards. Contrary to expectation from the BDC results, *L. gibbus* does not occupy a noticeably intermediate position between the two families. It is the closest liparid to the Cyclopteridae, but it definitely clusters with the other liparid species. These MDS results are also quite good, with a 3D stress of 0.116 and a minimal stress of 0.091 at five dimensions. Consequently, it is unlikely that *L. gibbus* would occupy any more of an intermediate position than it already does if additional scaling dimensions were considered.

The results seem to unambiguously support the classification of the Liparidae as a holobaramin. Continuity with Liparidae is supported by significant, positive BDC and clustering in MDS, and discontinuity between liparids and cyclopterids is supported by significant, negative BDC and separation in MDS. The only drawback to this conclusion is the small sampling of liparid genera (roughly a third: 9 of 29) and the significant, positive BDC between the outgroup and *L. gibbus*. Additional genera can only be added by further liparid research, and the BDC between cyclopterids and *L. gibbus* does not seem to be supported or explained by the MDS results. Given the data at hand, Liparidae seems to be a holobaramin.

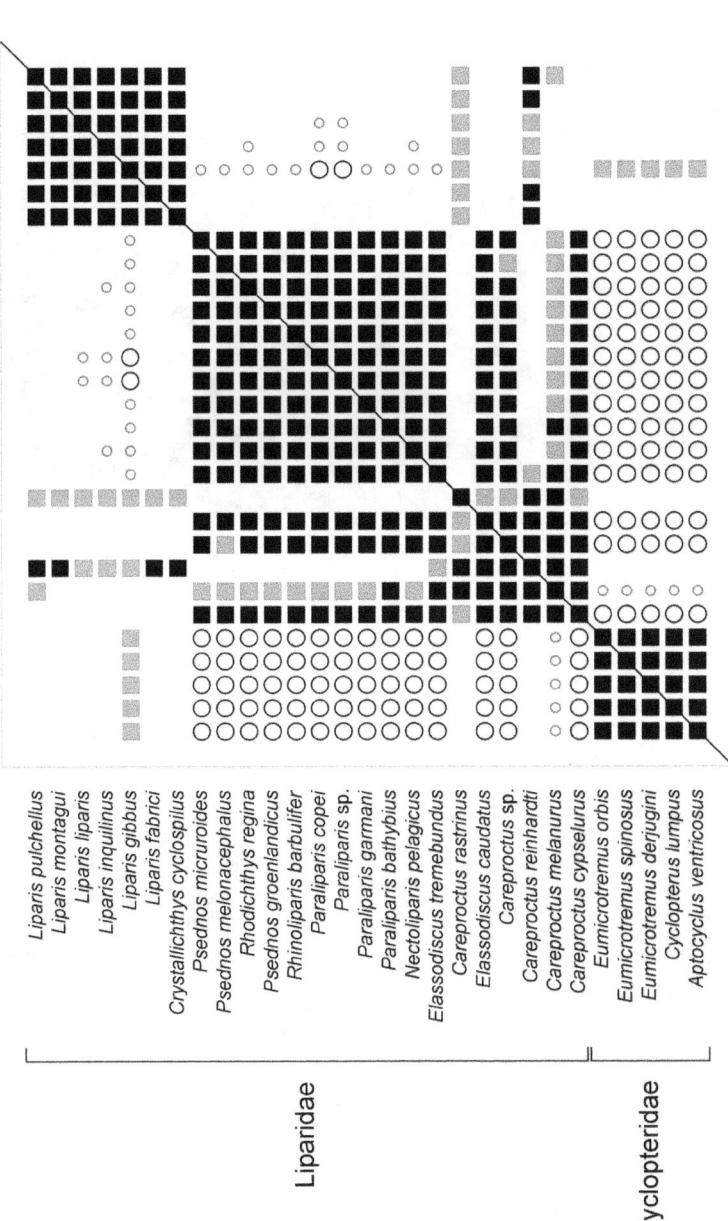

Figure 54. BDC bootstrap results for Liparidae, as calculated by BDISTMDS (relevance cutoff 0.95). Closed square indicate significant, positive BDC; open circles indicate significant, negative BDC. Black symbols indicate bootstrap values > 90% in a sample of 100 pseudoreplicates. Grey symbols represent bootstrap values ≤ 90%.

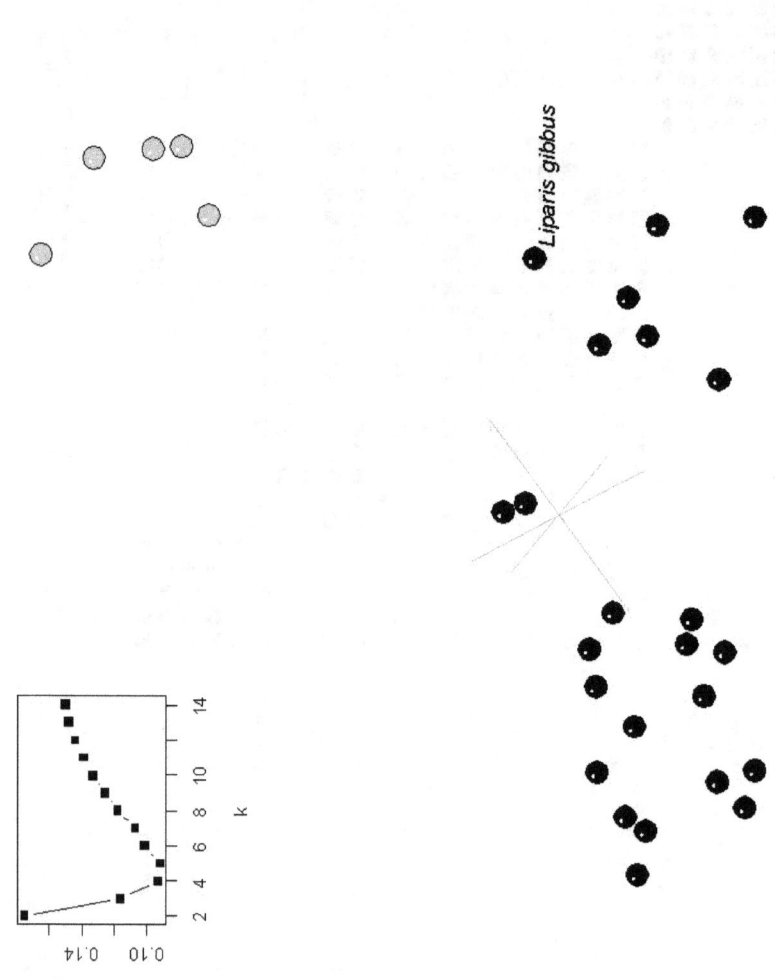

Figure 55. Three-dimensional MDS applied to Liparidae baraminic distances and the stress of *k*-dimensional MDS on the same baraminic distance matrix plotted as a function of the number of dimensions (*k*). Liparids are shown in black and outgroup taxa in gray.

## 2.26. Gasterosteidae (Vertebrata: Osteichthyes: Gasterosteiformes)

Dataset published by McLennan and Mattern (2001)

| | |
|---|---|
| Characters in published dataset: | 140 |
| Taxa in published dataset: | 7 |
| Character relevance cutoff: | 0.95 |
| Characters used to calculate BD: | 81 |
| Taxa used in BDC and MDS analysis: | 7 |
| Stress for 3D MDS: | 0.059 |
| $k_{min}$: | 4 |
| Median bootstrap value: | 92 |
| $F_{90}$: | 0.571 |

There are about seven species of sticklebacks (Gasterosteidae) in five genera: *Spinachia, Apeltes, Culaea, Pungitius,* and *Gasterosteus*. Named for the spines projecting from their backs, sticklebacks have bony armor and are found in temperate climates. Some species, such as *C. inconstans* are freshwater while others such as *A. quadracus* live in marine environments. McLennan and Mattern (2001) assessed the phylogeny of sticklebacks from 47 behavioral and 93 morphological characters (four "ambiguous" characters were omitted from their analysis). Their results showed that the geographically restricted *Spinachia* and *Apeltes* were basal to the more widespread *Pungitius* and *Gasterosteus*.

All of McLennan and Mattern's seven taxa were retained for baraminic distance calculations. The default character relevance cutoff of 0.95 required that character states be known for all seven taxa, which 89 characters met (34 morphological, 47 behavioral). The BDC results supported the clustering of only three pairs of taxa (Figure 56). *Spinachia* was correlated with the composite outgroup, the two species of *Gasterosteus* clustered together, and *Pungitius* and *Culaea* were positively correlated. Significant negative BDC was limited to comparisons of *Gasterosteus* species with the *Spinachia*/outgroup cluster and comparisons of the outgroup with the *Pungitius*/*Culaea* cluster. Bootstrap values for all correlations were quite good (median 92%).

Multidimensional scaling of these baraminic distances produced excellent results (3D stress 0.059). The distribution of taxa in MDS is quite diffuse and closely follows the phylogenetic results of McLennan and Mattern (Figure 57). The nearest taxon to the outgroup is *Spinachia*; as in the phylogeny, *Spinachia* is the most basal stickleback in the Gasterosteidae clade. *Apeltes* appears between *Spinachia* and the remaining sticklebacks, as in the branching order of the phylogeny. The remaining clusters of *Gasterosteus* species and *Pungitius/Culaea* are each monophyletic in the phylogeny.

It is difficult to assess the baraminic status of the sticklebacks from these results. The characters seem to be quite holistic (including morphological and behavioral) but the outgroup is a composite of outgroup taxa and does not represent a real fish. The clustering pattern of the gasterosteids is quite diffuse, but it is uncertain whether this is due to lack of a real outgroup, unconscious bias in character selection, or real discontinuities within the family. Future studies could resolve these questions by examining a slightly broader set of actual outgroup taxa, sampling the tubesnouts of family Aulorhynchidae as the putative sister taxon to the gasterosteids. The present study is inconclusive.

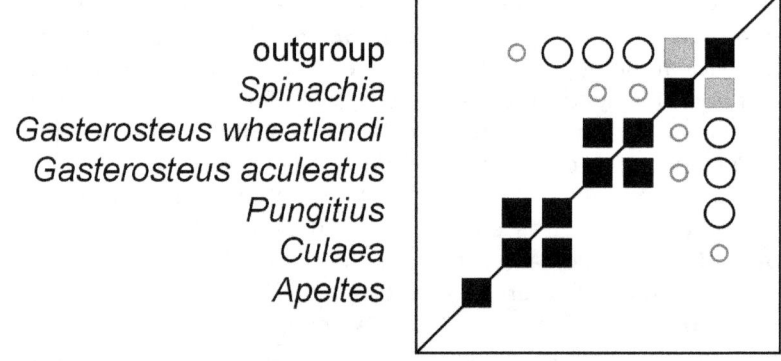

Figure 56. BDC bootstrap results for Gasterosteidae, as calculated by BDISTMDS (relevance cutoff 0.95). Closed square indicate significant, positive BDC; open circles indicate significant, negative BDC. Black symbols indicate bootstrap values >90% in a sample of 100 pseudoreplicates. Grey symbols represent bootstrap values ≤90%.

# Animals

101

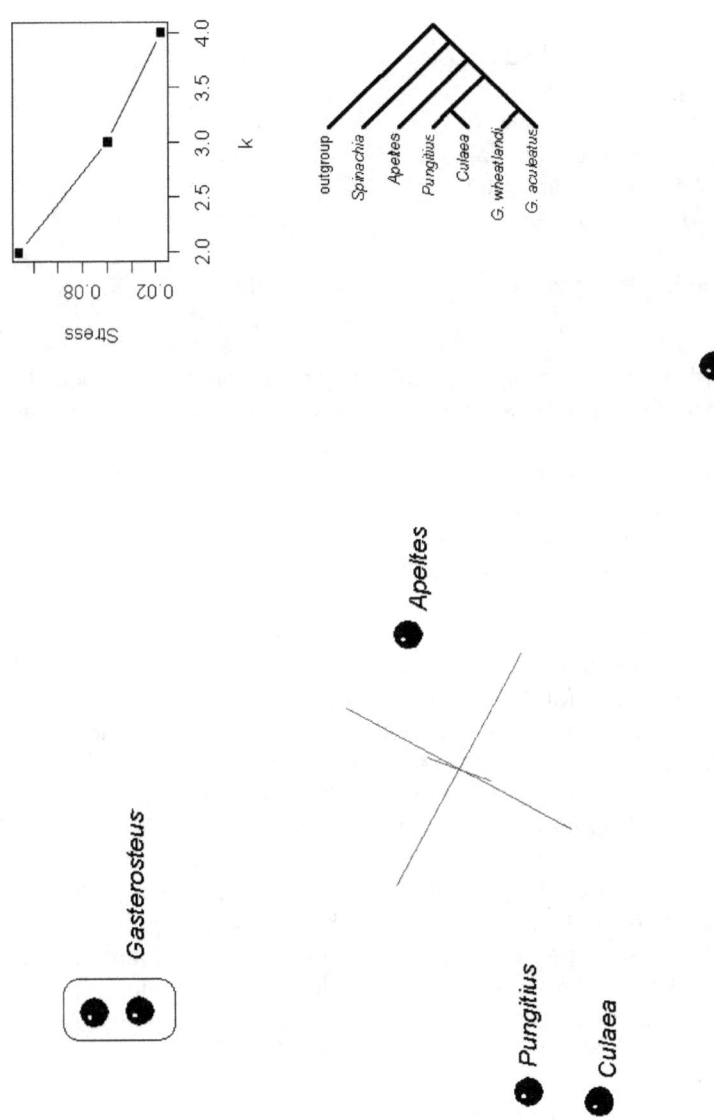

Figure 57. Three-dimensional MDS applied to Gasterosteidae baraminic distances and the stress of $k$-dimensional MDS on the same baraminic distance matrix plotted as a function of the number of dimensions ($k$). Also shown is the phylogeny from McLennan and Mattern (2001). Gasterosteids are shown in black and the outgroup in gray.

### 2.27. Stomiidae (Vertebrata: Osteichthyes: Stomiiformes)

Dataset published by Fink (1985)

| | |
|---|---|
| Characters in published dataset: | 323 |
| Taxa in published dataset: | 27 |
| Character relevance cutoff: | 0.95 |
| Characters used to calculate BD: | 305 |
| Taxa used in BDC and MDS analysis: | 27 |
| Stress for 3D MDS: | 0.181 |
| $k_{min}$: | 14 |
| Median bootstrap value: | 98 |
| $F_{90}$: | 0.618 |

Stomiidae are commonly referred to as dragonfish or viperfish, deep sea fish which use bioluminescence to attract prey. Some species have photophores on their bellies, while others have specialized barbels that terminate in luminescent organs. There are well over 200 stomiid species in around 27 genera. Fink included 25 genera and one composite outgroup in his 1985 phylogenetic study. His dataset consisted of 323 characters from 190 cranial, 118 postcranial, 8 light organ, and 7 miscellaneous (integumentary and life history) attributes.

Of the 323 characters, 305 were used for baraminic distance calculation after filtering at a character relevance cutoff of 0.95. Deleted were three cranial, fourteen postcranial, and one morphological characters (mental barbel presence/absence). Two groups are apparent in the BDC results, one of which corresponds to an unnamed clade in Fink's phylogeny (Group 2 in Figure 58). Within each group, significant, positive BDC is observed for most but not all taxon pairs. Between the groups, only significant, negative BDC is observed. Bootstrap values for most correlations are high (median 98%). The composite outgroup correlates positively with 12 of the 14 members of stomiid group 1.

The MDS results are poor with a 3D stress of 0.18 and a minimum stress of 0.014 at 14 dimensions. Nevertheless, the 3D pattern reveals two clusters of taxa that correspond to the two groups identified in the BDC results (Figure 59). The outgroup is near the edge of group 1 but definitely a part of that cluster of taxa.

Without a true set of outgroup taxa, it is impossible to verify the discontinuity within the stomiid family that is implied by significant, negative BDC between the two groups. Given the present data, however, there seems to be two holobaramins of stomiids. The first consists of the

14 genera labeled group 1 in Figure 58. Since the outgroup is only a representative composite and not a real organism, it is excluded from being a member of a holobaramin by definition; although, the significant, positive BDC between the outgroup and the stomiids might imply that continuity is shared between members of group 1 and other fish. In such a case, group 1 would actually be a polybaramin rather than a holobaramin. The second holobaramin consists of the remaining 12 genera labeled group 2 in Figure 58. This group contains no outgroup taxa and is therefore the more certain of the two holobaramins. Each holobaramin is united by continuity among its genera (as evidenced by significant, positive BDC) and separated from other taxa by discontinuity (as evidenced by significant, negative BDC).

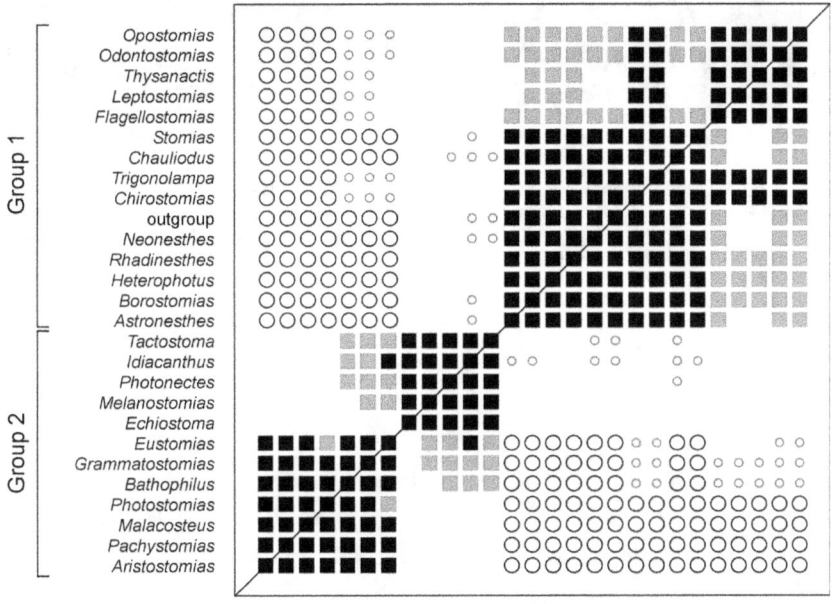

Figure 58. BDC bootstrap results for Stomiidae, as calculated by BDISTMDS (relevance cutoff 0.95). Closed square indicate significant, positive BDC; open circles indicate significant, negative BDC. Black symbols indicate bootstrap values >90% in a sample of 100 pseudoreplicates. Grey symbols represent bootstrap values ≤90%.

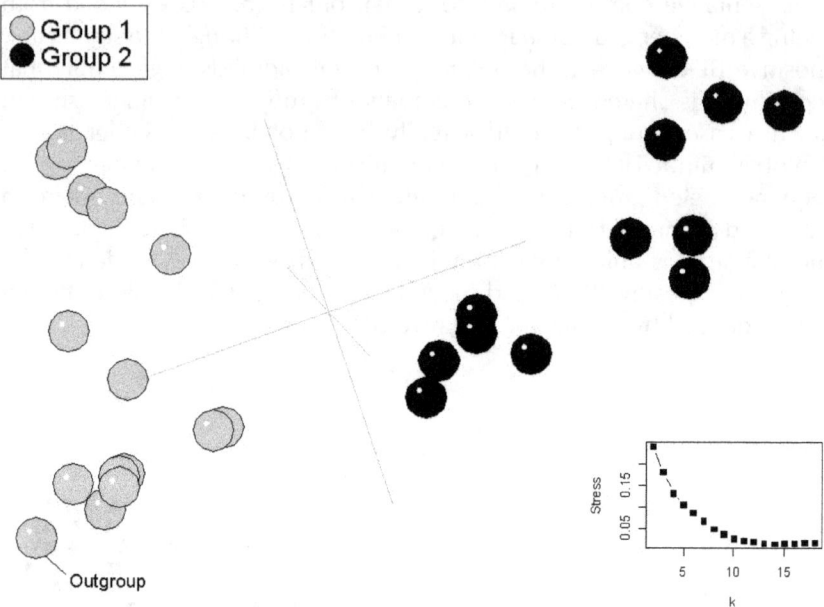

Figure 59. Three-dimensional MDS applied to Stomiidae baraminic distances and the stress of *k*-dimensional MDS on the same baraminic distance matrix plotted as a function of the number of dimensions (*k*). The two groups of stomiids seen in the BDC results (Figure 58) are shown in black and gray.

## 2.28. Epimeriidae and Iphimediidae (Arthropoda: Malacostraca: Amphipoda)

Dataset published by Lörz and Held (2004)

| | |
|---|---|
| Characters in published dataset: | 98 |
| Taxa in published dataset: | 16 |
| Character relevance cutoff: | 0.95 |
| Characters used to calculate BD: | 98 |
| Taxa used in BDC and MDS analysis: | 16 |
| Stress for 3D MDS: | 0.124 |
| $k_{min}$: | 7 |
| Median bootstrap value: | 93 |
| $F_{90}$: | 0.583 |

The amphipod families Epimeriidae and Iphimediidae are tiny, benthic, shrimp-like crustaceans. There are 25 species in six genera in Epimeriidae and 48 species in 13 genera in Iphimediidae. They both prefer polar waters and are found near both poles. The two families are classified in the superfamily Iphimedoidea. The phylogenetic study of Lörz and Held (2004) only partially sampled the taxa found in the southern hemisphere: six species of *Epimeria* (the largest of the Epimeriidae genera) and eight species from four genera of Iphimediidae. They used both mitochondrial DNA and morphology in their analysis. For the morphological phylogeny, they utilized 98 characters from sixteen taxa (14 ingroups and two outgroups *Eusirus* and *Monoculodes*). The characters sampled exoskeletal attributes from the entire body.

All characters and taxa were used to calculate baraminic distances. Three groups are apparent in the BDC analysis (Figure 60), corresponding to the Epimeriidae, Iphimediidae, and outgroups. Within each group all taxon pairs have significant, positive BDC except one: the BDC for iphimediids *Ghathiphimedia mandibularis* and *Echiniphimedia hodgsoni* is not statistically significant. Between the Epimeriidae and Iphimediidae, only significant, negative BDC is observed, but between the outgroup and ingroup taxa, only three taxon pairs have significant BDC. *Monoculodes* is positively correlated with the epimeriids *E. robusta* and *E. georgia* and negatively correlated with the iphimediid *Echiniphimedia waegeli*. Bootstrap values are fairly high (median 93%) especially between the two families; 35 of the 48 epimeriid/iphimediid taxon pairs have bootstrap values >90%.

The MDS shows the two amphipod families as elongate, parallel clusters with the outgroups adjacent to one end (Figure 61). Although *Monoculodes* was positively correlated with the Epimeriidae, it does not appear to be part of that cluster. It is interesting to note that the diversity of the epimeriid species of one genus is at least as much as the four genera of iphimediids, as approximated by the size of the taxic distribution in MDS.

The main drawback to this analysis is the limited sampling of the diversity of each family. Only one of six epimeriid genera and four of thirteen iphimediid genera are represented. Any baraminological conclusions we draw should be balanced by recognizing the need to have a better sampling of each family. The BDC and MDS results do support the presence of a discontinuity separating the Epimeriidae and Iphimediidae. The significant, negative BDC between the two families and their separation in MDS supports the inference of discontinuity. Likewise, the significant, positive BDC observed within each family supports their monobaraminic status. The significant, positive BDC between *Monoculodes* and two epimeriid species seems to be spurious, since there is a definite separation between the taxa in MDS. Consequently, pending further analysis with a greater sampling of taxa, Epimeriidae and Iphimediidae can be classified as holobaramins.

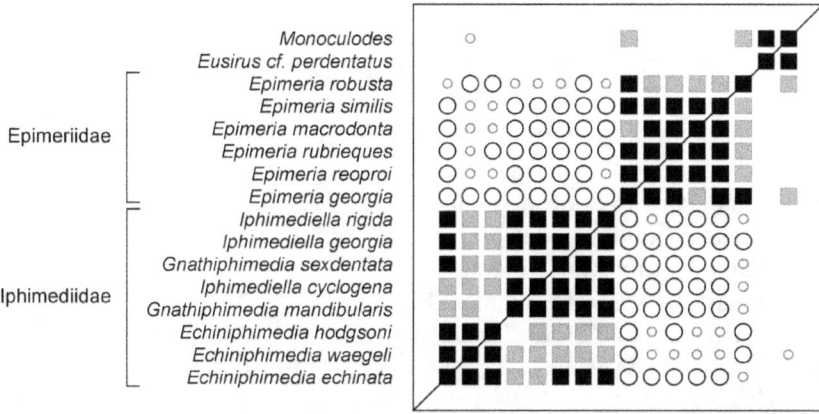

Figure 60. BDC bootstrap results for Epimeriidae and Iphimediidae, as calculated by BDISTMDS (relevance cutoff 0.95). Closed square indicate significant, positive BDC; open circles indicate significant, negative BDC. Black symbols indicate bootstrap values >90% in a sample of 100 pseudoreplicates. Grey symbols represent bootstrap values ≤90%.

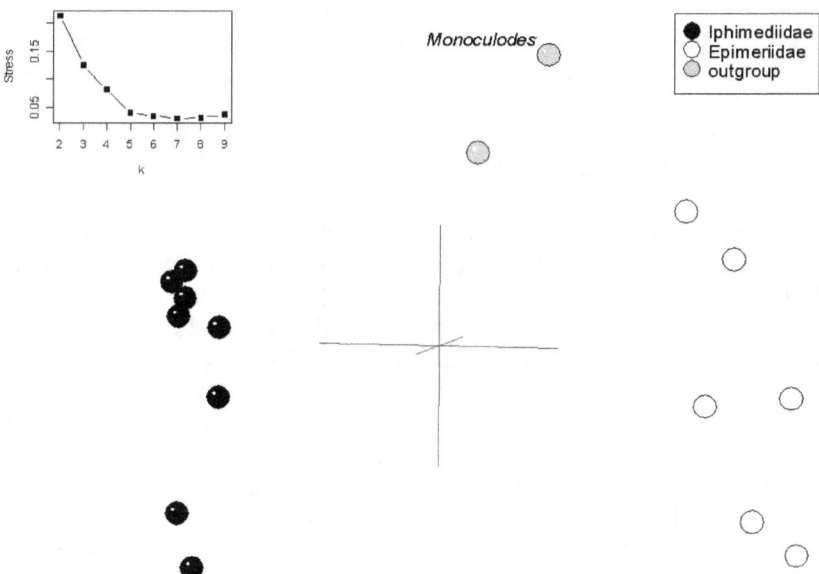

Figure 61. Three-dimensional MDS applied to Epimeriidae and Iphimediidae baraminic distances and the stress of k-dimensional MDS on the same baraminic distance matrix plotted as a function of the number of dimensions (k). Iphimediids are shown in black, epimeriids in white, and the outgroup taxa in gray.

### 2.29. Pholcidae (Arthropoda: Arachnida: Araneae)

Dataset published by Bruvo-Mađarić et al. (2005)

| | |
|---|---|
| Characters in published dataset: | 45 |
| Taxa in published dataset: | 34 |
| Character relevance cutoff: | 0.95 |
| Characters used to calculate BD: | 35 |
| Taxa used in BDC and MDS analysis: | 34 |
| Stress for 3D MDS: | 0.137 |
| $k_{min}$: | 6 |
| Median bootstrap value: | 62 |
| $F_{90}$: | 0.25 |

The house spiders of the family Pholcidae are web-weavers found worldwide. Huber (2000) listed 64 pholcid genera, which he placed in four major clades based on a morphological phylogeny. Three of his clades correspond roughly to previously recognized subfamilies: Pholcinae, Holocneminae, and Ninetinae, and the final clade consisted entirely of New World species. Bruvo-Mađarić et al. (2005) recently analyzed pholcid relationships using 45 characters from 31 pholcids and three outgroup taxa. The characters represent both exoskeletal and soft tissue attributes. The pholcid taxa sampled 23 of the 64 described genera representing all four of Huber's (2000) major clades. Using only the morphological data, Bruvo-Mađarić et al. found that Pholcidae is monophyletic and that the monophyly of each major group of pholcids is also supported except the ninetines which was represented by a single taxon.

All taxa and 35 characters (omitted: 17-18, 21-22, 25, 31-33, 35, 38) were used to calculate baraminic distances. BDC results revealed three groups: the outgroup taxa, the New World clade, and the remaining pholcids (Figure 62). Within the New World clade, all taxon pairs have significant, positive BDC with bootstrap values >90%. Within the group of remaining pholcids, only eight of 153 taxon pairs had no significant BDC, and all eight involved species of the genus *Physocyclus*. The remaining 145 taxon pairs of the "other pholcid" group had significant, positive BDC. The two pholcid groups had significant overlap. For example, both *Physocyclus* species (from the "other pholcid" group) are positively correlated with twelve of the thirteen New World pholcid species. Similarly, *Priscula* sp. is positively correlated with seven of the eighteen New World clade pholcids. The outgroup taxa are not positively correlated with any of the pholcids, but 45 of the 93 outgroup/pholcid taxon pairs have significant, negative BDC.

In the MDS results, two clear groups are visible, corresponding to the Pholcidae and the outgroup taxa (Figure 63). The pholcids are distributed in a bi-lobed cluster, which accounts for the two groups of pholcids apparent in the BDC results.

These results strongly indicate that the pholcid spiders are members of a single holobaramin. Significant, positive BDC and the close MDS clustering support the continuity of the pholcid species. Significant, negative BDC between the pholcids and outgroup taxa and their separation in MDS supports the discontinuity between the pholcids and the outgroups. The dataset is also holistic, and the bootstrap results for the positive BDC indicate that these correlations are robust to random perturbations in the dataset. Given these factors, Pholcidae is a holobaramin, the first holobaramin of spiders ever to be identified.

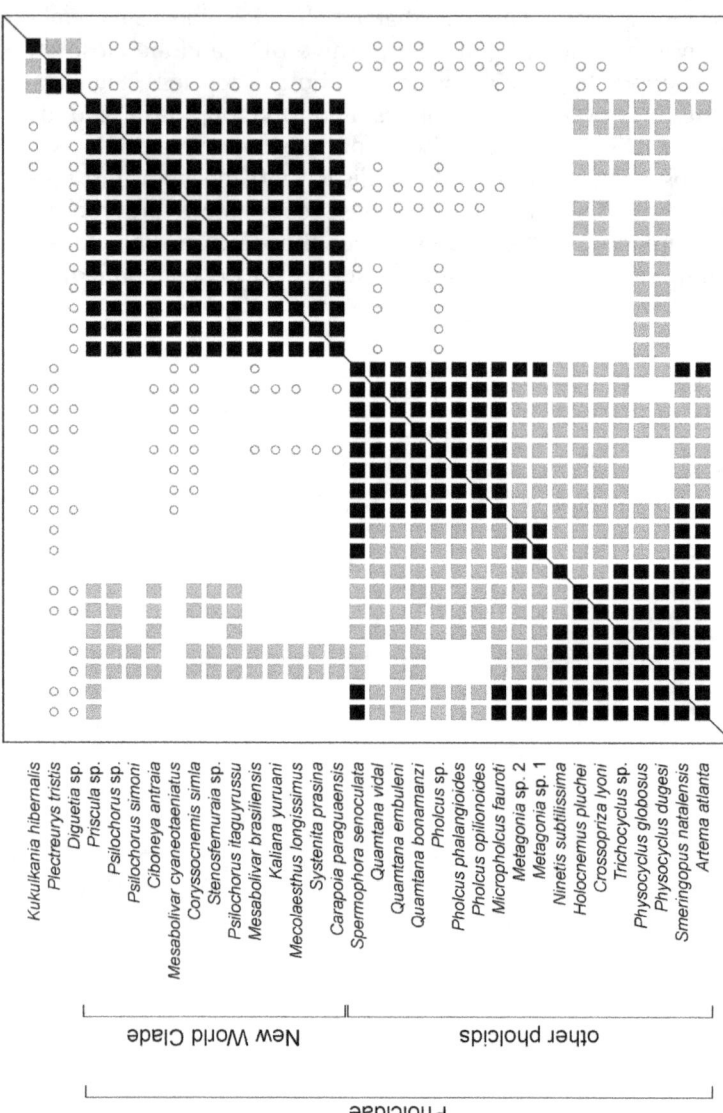

Figure 62. BDC bootstrap results for Pholcidae, as calculated by BDISTMDS (relevance cutoff 0.95). Closed square indicate significant, positive BDC; open circles indicate significant, negative BDC. Black symbols indicate bootstrap values >90% in a sample of 100 pseudoreplicates. Grey symbols represent bootstrap values ≤90%.

# Animals

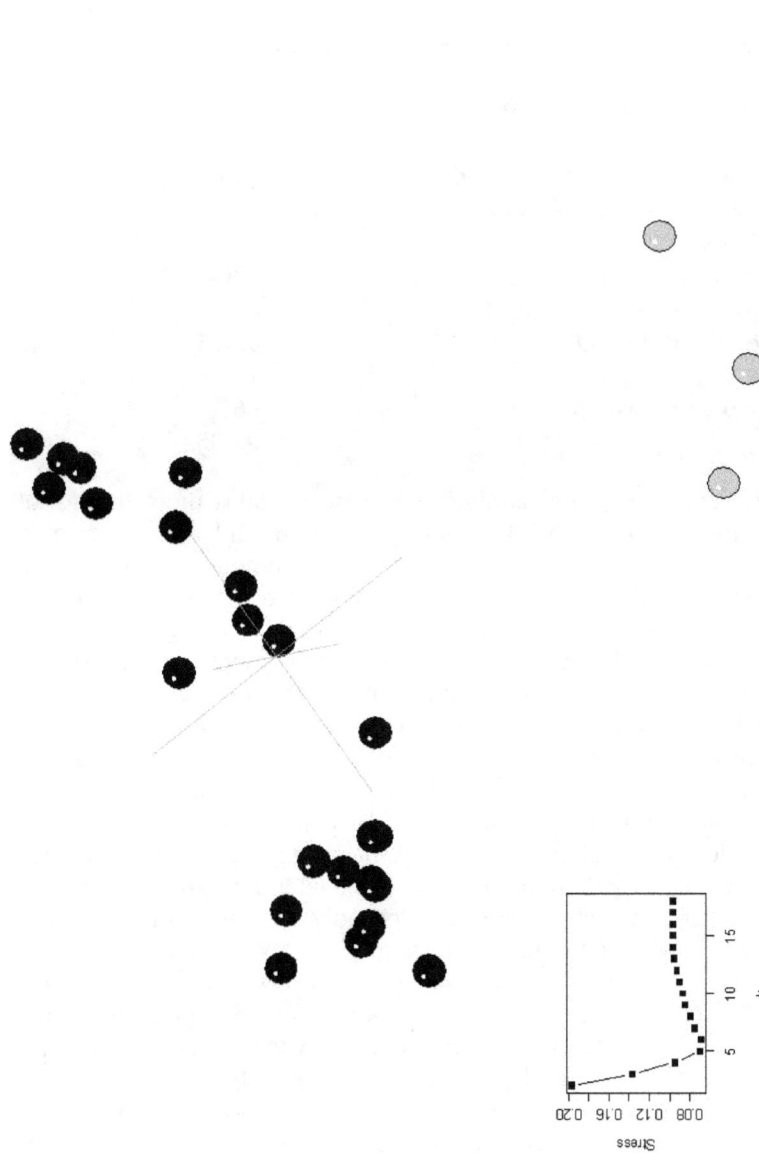

Figure 63. Three-dimensional MDS applied to Pholcidae baraminic distances and the stress of $k$-dimensional MDS on the same baraminic distance matrix plotted as a function of the number of dimensions ($k$). Pholcids are shown in black and outgroup taxa in gray.

## 2.30. Theridiidae (Arthropoda: Arachnida: Araneae)

Dataset published by Agnarsson (2004)

| | |
|---|---|
| Characters in published dataset: | 242 |
| Taxa in published dataset: | 62 |
| Character relevance cutoff: | 0.95 |
| Characters used to calculate BD: | 220 |
| Taxa used in BDC and MDS analysis: | 62 |
| Stress for 3D MDS: | 0.305 |
| $k_{min}$: | 13 |
| Median bootstrap value: | 96 |
| $F_{90}$: | 0.608 |

The cobweb spider family Theridiidae contains more than 2000 species in approximately 80 genera. Theridiid spiders are extremely diverse, with extremely toxic species (the famous widow spiders *Latrodectus*), solitary myrmecophagous species (the peculiar ant-mimicking *Anatea*), social spiders that build massive webs (the southeastern social spider *Anelosimus*), and the kleptoparasitic spiders that live on other species' webs (dewdrop spiders *Argyrodes*). Agnarsson (2004) recognized six clades within the monophyletic Theridiidae: Hadrotarsinae, Latrodectrinae, Spintharinae, Pholcommatinae, Argyrodinae, and Theridiinae. Four of these subfamilies are also clades in the molecular phylogeny of Arnedo et al. (2004), although their phylogenetic relations differ. Agnarsson's (2004) morphological phylogeny was based on 242 mostly exoskeletal but also soft tissue and natural history characters. He included only 53 theridiid taxa from 32 genera along with eight outgroups.

Since Agnarsson's dataset was very complete, only 22 characters (74, 108, 171, 217, 224-235, 237-242) were omitted from the baraminic distance calculations. There are approximately three groups in the BDC results, two of which overlap (Figure 64). The first group is composed of the eight outgroup taxa, and the second group corresponds to the Hadrotarsinae and two genera of pholcommatines, *Phoroncidia* and *Cerocida*. The third group contains the remaining theridiids. Within each group significant, positive BDC is frequent and high (>90%) bootstrap values are common. The two theridiid groups overlap. The hadrotarsine *Dipoena* is positively correlated with two latrodectines and two spintharines. *Phoroncidia* and *Cerocida* correlate positively with three and six other pholcommatines in the main theridiid group, respectively. Otherwise, the smaller theridiid group is negatively correlated with 32 of the main theridiid group. The outgroup taxa are negatively correlated with nearly every one of the main group of theridiid taxa, and most of these correlations have high bootstrap values (>90%). Two of the smaller theridiid group (*Emertonella* and *Cerocida*) are positively correlated with the outgroup nesticid *Eidmanella*.

MDS results support the presence of three groups, though the 3D stress is high (0.305). The smaller theridiid group could be seen as a lobe of the main theridiid group rather than as a separate group (Figure 65). The outgroup taxa are definitely separate from the main theridiid cluster(s). Nesticids (*Nesticus* and *Eidmanella*) are the closest outgroup taxa to the theridiids, coming nearest to the taxa of the smaller theridiid group.

From the BDC results alone, Theridiidae could be a holobaramin based on the significant, positive BDC within the group and significant, negative BDC between the theridiids and the outgroups. The smaller theridiid group could be part of the main theridiid group or part of the outgroup. In the case of the Poaceae (Wood 2002), three taxa, *Streptochaeta*, *Anommochloa*, and *Pharus*, were initially classified as inconclusive based on BDC analysis alone. Subsequent MDS analysis supported the grouping of *Pharus* with the grasses and *Streptochaeta* and *Anommochloa* with the outgroup taxa (Wood 2005b). In this case, the MDS results are not entirely conclusive. It is possible to view the smaller group of theridiids as part of the main Theridiidae, which would seem to be warranted by the significant, positive BDC observed between the two theridiid groups. Further, the significant, positive BDC between the smaller theridiid group and the outgroup is limited to two instances involving only three taxa. Based on the MDS results, these taxa are adjacent but separated into two different clusters. Considering all these factors, it would seem that the entire family Theridiidae is a single holobaramin. Since fossils of basal theridiid subfamilies are known from Baltic amber (Eocene) and additional subfamilies are found in Dominican amber (Miocene) (Marusik and Penney 2004), it would appear that modern theridiid species arose from post-Flood diversification. It is not immediately apparent how many Flood survivors originated this diversification.

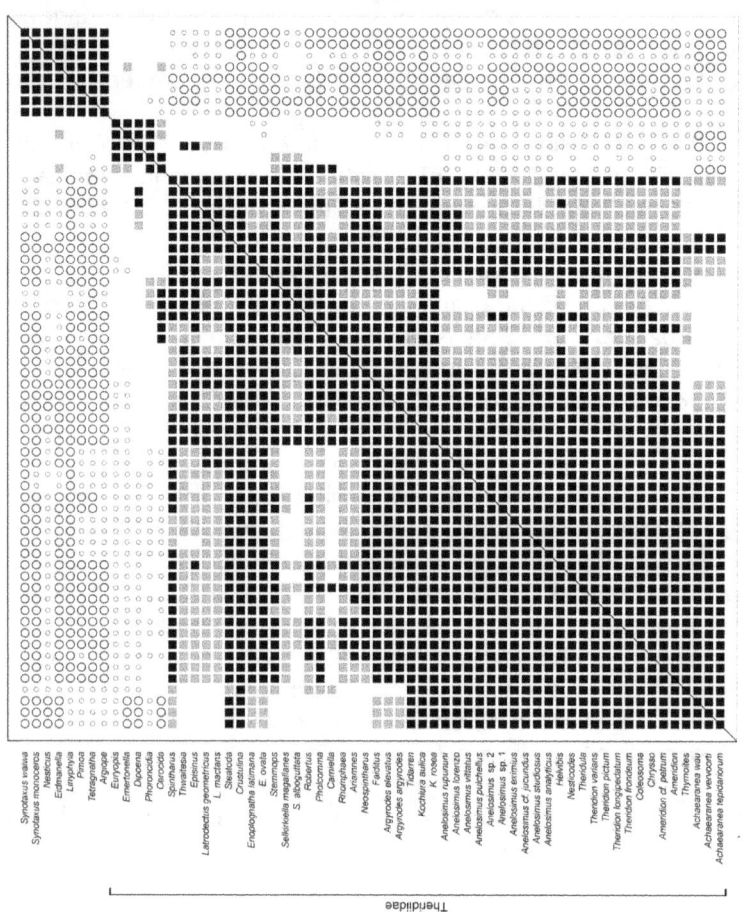

Figure 64. BDC bootstrap results for Theridiidae, as calculated by BDISTMDS (relevance cutoff 0.95). Closed square indicate significant, positive BDC; open circles indicate significant, negative BDC. Black symbols indicate bootstrap values >90% in a sample of 100 pseudoreplicates. Grey symbols represent bootstrap values ≤90%.

# Animals

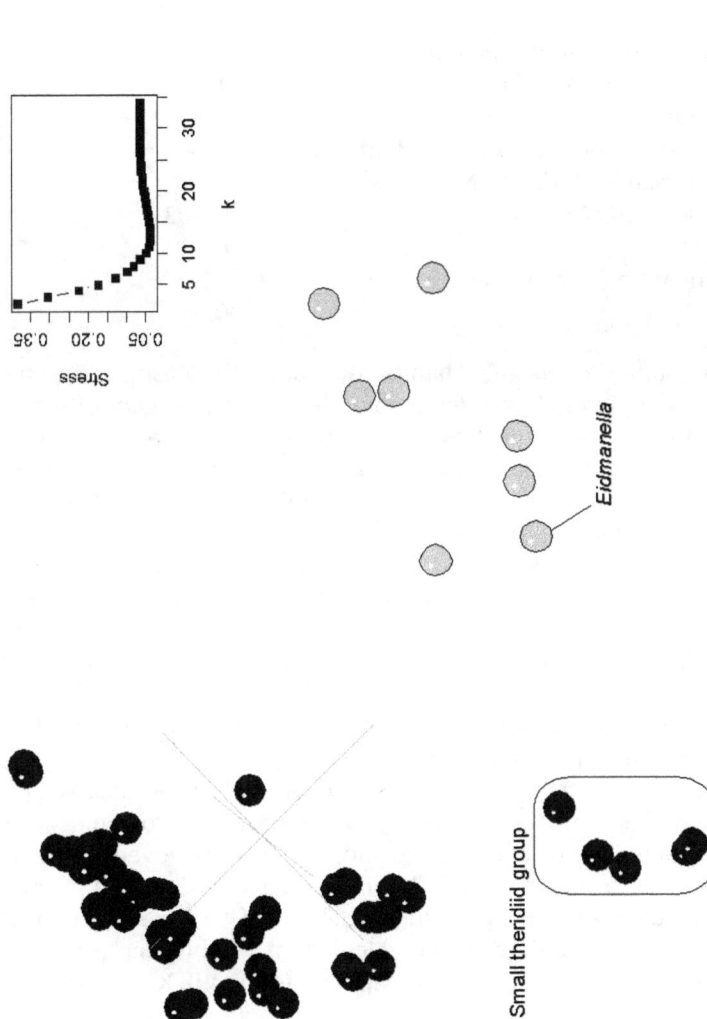

Figure 65. Three-dimensional MDS applied to Theridiidae baraminic distances and the stress of $k$-dimensional MDS on the same baraminic distance matrix plotted as a function of the number of dimensions ($k$). Theridiids are shown in black, outgroup taxa in gray.

## 2.31. Sarcoptidae (Arthropoda: Arachnida: Astigmata)

Dataset published by Klompen (1992)

| | |
|---|---|
| Characters in published dataset: | 215 |
| Taxa in published dataset: | 106 |
| Character relevance cutoff: | 0.9 |
| Characters used to calculate BD: | 114 |
| Taxa used in BDC and MDS analysis: | 91 |
| Stress for 3D MDS: | 0.103 |
| $k_{min}$: | 6 |
| Median bootstrap value: | 96 |
| $F_{90}$: | 0.61 |

Sarcoptidae is a mite family of obligate parasites, the most notorious of which is *Sarcoptes scabiei*, the causative agent of scabies. Only mammals serve as hosts to sarcoptids, primarily marsupials, rodents, insectivores, bats, and primates. Klompen's (1992) is one of few phylogenetic studies of the sarcoptids and is based only on morphology. No molecular phylogenies of the family have been published as of this writing. Klompen recognized three monophyletic subfamilies: Diabolicoptinae, consisting of two genera and three species; Sarcoptinae, comprised of four genera and eight species; and the largest Teinocoptinae, encompassing nine genera and 106 species, for a total of 15 genera and 117 species. Klompen's phylogenetic analysis was based on 215 morphological characters, primarily (but not exclusively) external cuticular characters, and 106 taxa.

Due to the incompleteness of the data, the eleven taxa with the lowest taxic relevance (0.367-0.526) were removed from the dataset prior to baraminic distance calculations. Also removed were four hypothetical ancestral taxa included in the data matrix. At a character relevance cutoff of 0.9, the reduced dataset of 91 taxa retained 114 characters for baraminic distance calculations, slightly more than half of the original set. Two groups are apparent in the BDC results (Figure 66). The first group corresponds to the *Cynopterocoptes/Teinocoptes* clade Klompen found in the subfamily Teinocoptinae. The other group consists of the remaining sarcoptid taxa and the outgroup family Rhynchoptidae. Within each group, significant, positive BDC is observed for most taxon pairs with bootstrap values generally >90%. Only five instances of significant, negative BDC are observed within the larger group, all involving *Chirnyssoides* (C.) *brasiliensis*. Between the groups, significant, negative BDC is common, typically with bootstrap values >90%. One taxon, *Cynopterocoptes haeneyi*, connects the two groups by correlating positively with 14 species of *Nycteridocoptes*.

The MDS results do not fully support the two separate groups observed in the BDC results (Figure 67). The taxa appear more as a diffuse cloud with two lobes, which correspond to the *Cynopterocoptes/Teinocoptes* group and the remaining taxa. *Cynopterocoptes haeneyi* occupies an intermediate position between the two groups, and *Nycteridocoptes* most closely approaches it from the other group. The outgroup Rhynchoptidae is part of the larger group of taxa and is closest to the sarcoptine *Trixacarus eliurus*.

These results could indicate one of two possible baraminological interpretations. The *Cynopterocodes/Teinocoptes* group could be interpreted as a holobaramin, separated from the remaining sarcoptids and rhyncoptids by discontinuity. This interpretation would be supported by the significant, positive BDC within the *Cynopterocoptes/Teinocoptes* group and the significant, negative BDC between the *Cynopterocoptes/Teinocoptes* group and the remaining taxa. Drawbacks to this interpretation include the lack of separate groups corresponding to the BDC groups in the MDS results and the signficiant, positive BDC between *Cynopterocoptes haeneyi* and *Nycteridocoptes* species (implying continuity). This suggests a second interpretation, namely that the present data are inconclusive for resolving the baraminic relationships. It is possible that the Rhyncoptidae could be part of the same holobaramin as the sarcoptids, or the present character sample could be truly inadequate for identifying the holobaramin. One future area of possible study would be to examine a set of real rhyncoptids instead of the composite Rhyncoptidae. Inclusion of real outgroup taxa could help to resolve the ingroup relationships.

Figure 66. BDC bootstrap results for Sarcoptidae, as calculated by BDISTMDS (relevance cutoff 0.9). Closed square indicate significant, positive BDC; open circles indicate significant, negative BDC. Black symbols indicate bootstrap values > 90% in a sample of 100 pseudoreplicates. Grey symbols represent bootstrap values ≤90%.

Animals 119

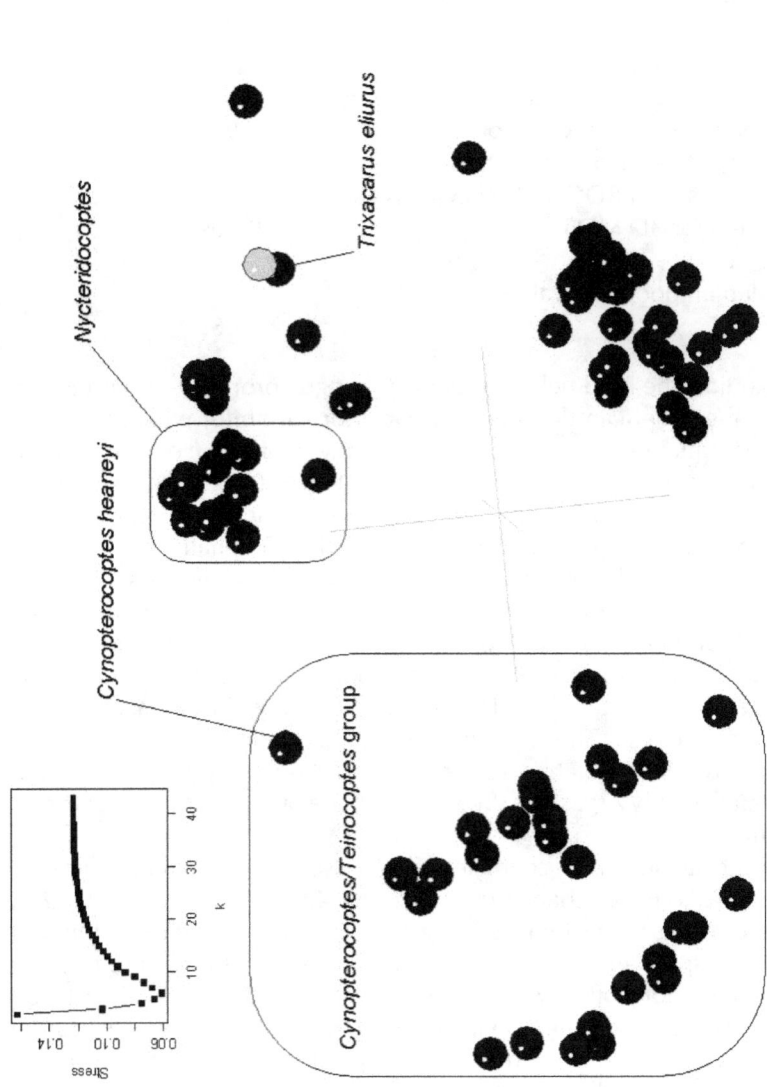

Figure 67. Three-dimensional MDS applied to Sarcoptidae baraminic distances and the stress of $k$-dimensional MDS on the same baraminic distance matrix plotted as a function of the number of dimensions ($k$). Sarcoptids are shown in black and the outgroup taxon in gray.

### 2.32. Ixodidae (Arthropoda: Arachnida: Ixodida)

Dataset published by Klompen et al. (2000)

| | |
|---|---|
| Characters in published dataset: | 125 |
| Taxa in published dataset: | 40 |
| Character relevance cutoff: | 0.95 |
| Characters used to calculate BD: | 115 |
| Taxa used in BDC and MDS analysis: | 40 |
| Stress for 3D MDS: | 0.109 |
| $k_{min}$: | 7 |
| Median bootstrap value: | 99 |
| $F_{90}$: | 0.665 |

Ixodids, the hard ticks, are obligate sanguivorous ectoparasites and include the American dog tick *Dermacentor variabilis*, which transmits Rocky Mountain spotted fever, as well as the deer tick *Ixodes dammini*, which transmits Lyme disease. Approximately 650 species of ticks have been described and placed in five subfamilies Ixodinae, Amblyomminae, Haemaphysalinae, Hyalomminae, and Rhiphicephalinae. Recent morphological and molecular phylogenetic analyses have not supported this subfamilial classification (Black and Piesman 1994; Klompen et al. 2000). Klompen et al.'s (2000) non-molecular dataset consisted of 125 characters scored for 40 taxa. The characters were a mix of soft-tissue and external characters from larvae and adults, together with six natural history characters. The taxa included 31 ixodids, representing seven genera and all the putative subfamilies. The outgroup taxa came from the soft tick family Argasidae (five taxa) and the other orders of superorder Parasitiformes (four taxa).

All taxa and 115 characters were used for baraminic distance calculations (omitted characters: 23-25, 38, 42, 44, 108, 115, 120-121). The BDC results reveal two distinct groups that correspond to the ingroup and outgroup taxa (Figure 68). Within each group, significant, positive BDC is common, often with bootstrap values >90%. All 279 taxon pairs of ingroup and outgroup taxa are negatively correlated, and 255 of these negative correlations have bootstrap values >90%. The Ixodidae appear to be divided into two subgroups, consisting of the genus *Ixodes* and the remaining ixodids. Within the subgroups, all taxon pairs have significant, positive BDC with bootstrap values >90%. While the subgroups are connected by 116 taxon pairs with significant, positive BDC, but only four of those have bootstrap values >90%.

Four groups are visible in the MDS results (Figure 69): *Ixodes*, the remaining ixodids, Argasidae, and the remaining outgroups. From the MDS alone, it does not appear that *Ixodes* and the remaining ixodids are part of the same cluster; nevertheless, the outgroup taxa are well-separated from the Ixodidae.

Although the two subgroups of Ixodidae are separated in the MDS, the extensive occurrence of significant, positive BDC between the two groups would imply that they belong together in the same monobaramin. Furthermore, the universal negative correlation between the outgroup taxa and the ixodids implies a discontinuity between the ixodids and other taxa. Taken together, then, these results strongly support the holobaraminic status of Ixodidae. Given the separation between the argasids and the other outgroup taxa, it is possible that the Argasidae too could be a holobaramin. Future studies should focus on verifying the cobaraminic status of *Ixodes* and the remaining ixodids and examining the baraminic status of Argasidae. Since ticks are obligate ectoparasites, they could make an interesting model system for understanding the origin of natural evil.

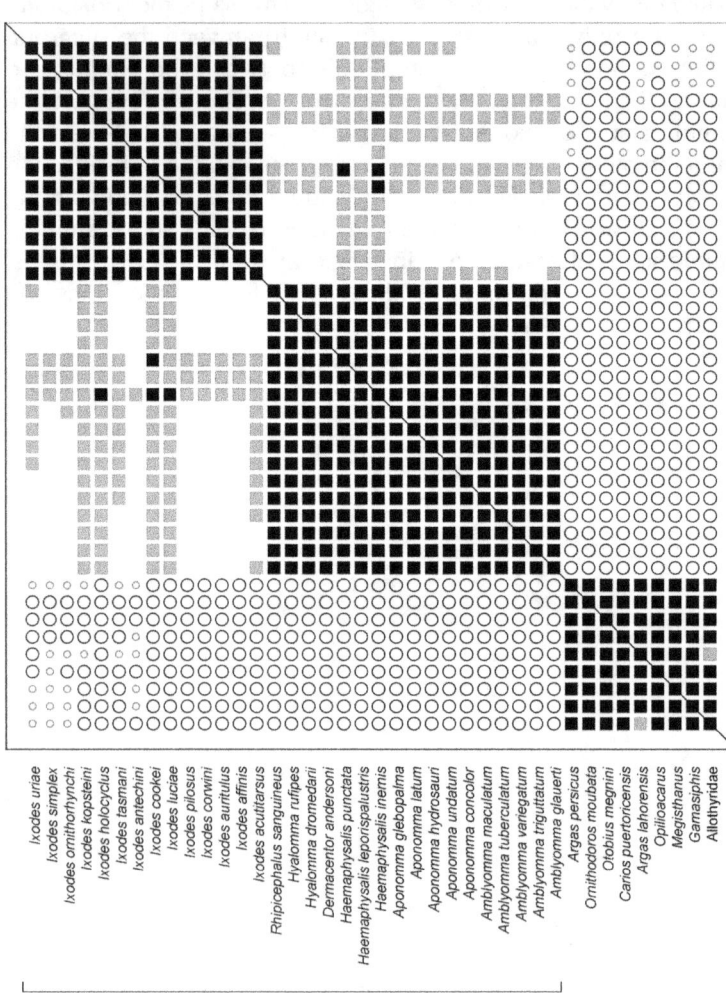

Figure 68. BDC bootstrap results for Ixodidae, as calculated by BDISTMDS (relevance cutoff 0.95). Closed square indicate significant, positive BDC; open circles indicate significant, negative BDC. Black symbols indicate bootstrap values >90% in a sample of 100 pseudoreplicates. Grey symbols represent bootstrap values ≤90%.

Animals 123

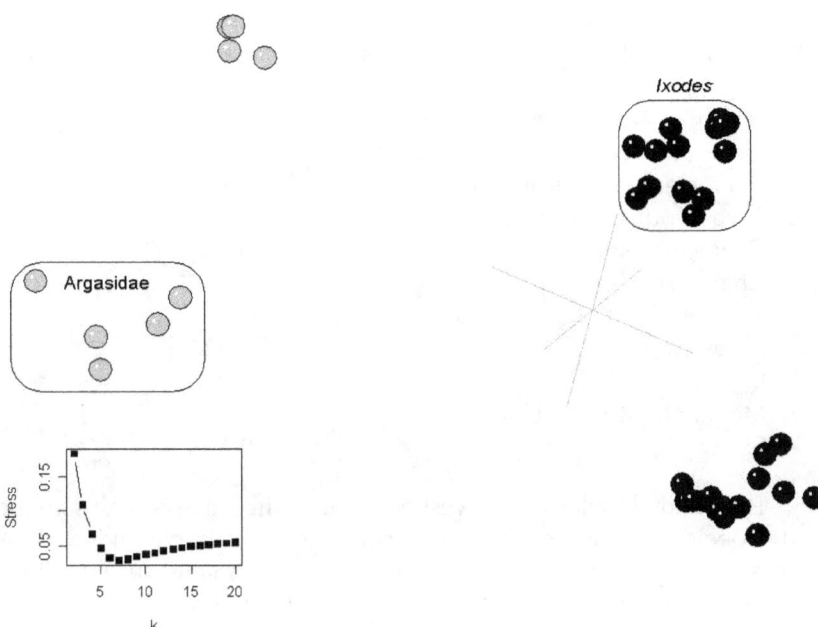

Figure 69. Three-dimensional MDS applied to Ixodidae baraminic distances and the stress of *k*-dimensional MDS on the same baraminic distance matrix plotted as a function of the number of dimensions (*k*). Ixodids are shown in black, outgroup taxa in gray.

## 2.33. Sironidae (Arthropoda: Arachnida: Opiliones)

Dataset published by de Bivort and Giribet (2004)

| | |
|---|---|
| Characters in published dataset: | 46 |
| Taxa in published dataset: | 27 |
| Character relevance cutoff: | 0.95 |
| Characters used to calculate BD: | 41 |
| Taxa used in BDC and MDS analysis: | 27 |
| Stress for 3D MDS: | 0.238 |
| $k_{min}$: | 6 |
| Median bootstrap value: | 75 |
| $F_{90}$: | 0.168 |

The daddy longlegs or harvestmen family Sironidae contains about thirty species in nine genera. Harvestmen are small arachnids whose bodies resemble mites. They live in leaf litter, chiefly in temperate regions of the northern hemisphere. They are common in Europe and North America, and one species *Suzukielus sauteri* occurs Japan. De Bivort and Giribet (2004) evaluated sironid phylogeny using a matrix of 46 characters and 27 taxa. Their characters were primarily exoskeletal, and the taxa consisted of 17 sironids and 10 outgroup taxa from five other harvestmen families (Petallidae, Ogoveidae, Stylocellidae, Neogoveidae, and Troglosironidae). Their phylogeny implied that the most basal of the sironids is the Japanese *Suzukielus*.

For baraminic distance calculations, all taxa and 36 of the 46 characters were used (omitted characters: 34, 43-46). There is significant, positive BDC between members of the Sironidae (Figure 70), but there appear to be two separate groups of sironids. The smaller group consists of the genera *Suzukielus*, *Parasiro*, and *Odontosiro*, genera which do not form a clade in de Bivort and Giribet's phylogenetic analysis. The remaining taxa comprise the larger group and are connected to the smaller by significant, positive BDC with *Suzukielus* and *Parasiro coiffaiti*. There are no cases of two sironids with significant, negative BDC. The outgroup taxa form a somewhat loosely associated group that are negatively correlated with sixteen of the sironids (*Odontosiro* is not significantly correlated with any outgroup taxa). Bootstrap values are poor for this dataset, with a median of only 75%. Highest bootstrap values occur in the larger sironid group (excluding *Suzukielus*, *Parasiro*, and *Odontosiro*). All significant, negative BDC has bootstrap values <90%.

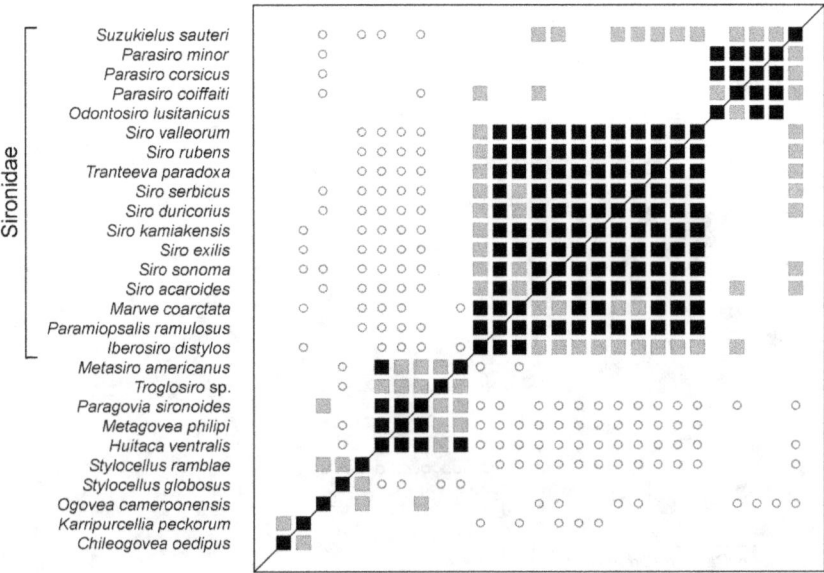

Figure 70. BDC bootstrap results for Sironidae, as calculated by BDISTMDS (relevance cutoff 0.95). Closed square indicate significant, positive BDC; open circles indicate significant, negative BDC. Black symbols indicate bootstrap values >90% in a sample of 100 pseudoreplicates. Grey symbols represent bootstrap values ≤90%.

The MDS results are unusually poor, with a 3D stress of 0.238. The sironids are distributed in a bilobed cluster with the outgroup taxa forming an arc around them (Figure 71). The two lobes of sironids correspond to the two sironid groups identified in the BDC analysis. The closest ingroup/outgroup pair are *Marwe coarctata* and *Chileogovea oedipus* (MDS distance 0.195), and these two are not significantly correlated in the BDC results. The widely spaced distribution of outgroup taxa probably explains why there is comparatively little significant, positive BDC within the outgroups.

The BDC results alone would support the holobaraminic classification of Sironidae. The significant, positive BDC between sironids indicates continuity within the group, and the significant, negative BDC with the outgroup taxa indicates a discontinuity. Surprisingly, the MDS results reveal an unusual distribution of taxa with a high degree of distortion of the baraminic distances (as indicated by the high 3D stress). Instead of forming two separate clusters of sironids and outgroups, the sironids form a bilobed cluster with the outgroups distributed in a kind of cloud around the sironids. Nevertheless, given the BDC results and pending further more detailed baraminological study of the harvestmen, the Sironidae appear to form a holobaramin.

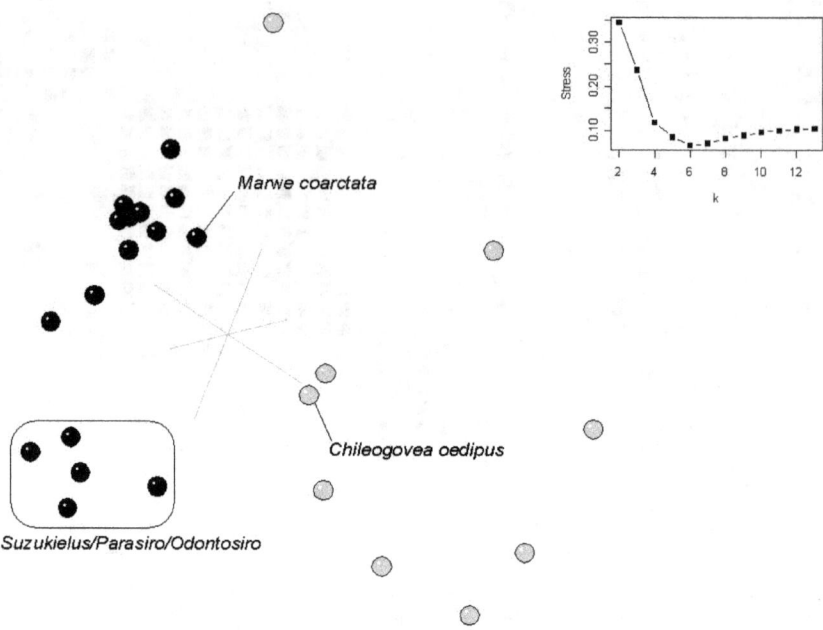

Figure 71. Three-dimensional MDS applied to Sironidae baraminic distances and the stress of *k*-dimensional MDS on the same baraminic distance matrix plotted as a function of the number of dimensions (*k*). Sironids are shown in black and outgroup taxa in gray.

## 2.34. Bothriuridae and other Scorpionoidea (Arthropoda: Arachnida: Scorpiones)

Dataset published by Prendini (2000)

| | |
|---|---|
| Characters in published dataset: | 115 |
| Taxa in published dataset: | 35 |
| Character relevance cutoff: | 0.95 |
| Characters used to calculate BD: | 114 |
| Taxa used in BDC and MDS analysis: | 71 |
| Stress for 3D MDS: | 0.200 |
| $k_{min}$: | 8 |
| Median bootstrap value: | 84 |
| $F_{90}$: | 0.462 |

According to Prendini's (2000) phylogenetic analysis, the scorpion superfamily Scorpionoidea is composed of seven families, forty genera, and more than 500 species. Among these scorpions are some of the largest in the world; the giant Indian black scorpion *Heterometrus swammerdami* (Scorpionidae) can reach 28 cm (11 inches). Though all scorpions produce venom for immobilizing their prey, typical prey are other arthropods and thus few scorpions present a direct threat to humans. Prendini studied scorpionoid phylogeny from 115 morphological characters, both exoskeletal and soft tissue. He sampled 71 species from 40 genera representing all scorpionoid families.

All taxa and 114 of Prendini's characters were used to calculate baraminic distances (omitted was character 79, straight or coiled testes). Three groups can be seen in the BDC results (Figure 72). The first group consists of the two outgroup taxa, *Chaerilus* and *Centruroides*, the second of the family Bothriuridae, and the third of the remaining scorpionoids. Within the Bothriuridae, all taxon pairs have significant, positive BDC with bootstrap values >90%. When bothriurids are compared to other scorpionoids, only significant, negative BDC is observed. Of the 1034 taxon pairs involving a bothriurid and another scorpionoid, 1005 are negative correlated, and 446 of those have bootstrap values >90%. The other scorpionoid group is more heterogeneous, with roughly two subgroups detectable. One subgroup corresponds to the families Ischnuridae, Urodacidae, and Heteroscorpionidae, and the other subgroup contains the Scorpionidae, Hemiscorpiidae, and Diplocentridae. Within each subgroup, most taxon pairs have signficiant, positive BDC with high (>90%) bootstrap values. Between the two subgroups, significant, positive BDC is also common, and no significant, negative BDC is present. Three taxa, corresponding to Prendini's family Hemiscorpiidae, are positively correlated with all members of both subgroups (but negatively correlated with bothriurids and the outgroups). The outgroup taxa are negatively correlated with members of the larger, multifamily scorpionoid group, but significant BDC (positive or negative) is not observed between the bothriurids and the outgroup taxa.

The MDS results are quite poor at three dimensions (3D stress 0.200), even though the minimal stress is only 0.032 at eight dimensions. The three dimensional pattern of taxa will consequently be distorted from its optimal distribution. Despite this distortion, the two scorpionoid groups are evident in 3D MDS, with the outgroup taxa between them (Figure 73). There does seem to be some separation between the larger families of the nonbothriurid scorpionoids (i.e. Scorpionidae, Ischnuridae, and Diplocentridrae), but the smaller families (Hemiscorpiidae, Heteroscorpionidae, and Urodacidae) bridge the gaps between them. The outgroup taxa seem surprisingly close to the bothriurids even though they do not share significant, positive BDC, but this adjacency may be due to the distortions of 3D MDS as indicated by the high stress.

Considering these results, Bothriuridae seems unquestionably a holobaramin, as do the remaining non-bothriurid scorpionoids. The taxa of each group are connected by significant, positive BDC and separated from taxa in the other group by significant, negative BDC. The MDS results confirm that these are separate groups with nonscorpionoids between them. Although the non-bothriurid scorpionoids appear to be a heterogeneous group, and possibly multiple holobaramins, the distribution of taxa in MDS and the BDC results do not support splitting the group into multiple holobaramins. Given this classification, the bothriurid holobaramin contains about 130 species, and the non-bothriurid scorpionoid holobaramin has around 400 species. Since all of these species are venomous, this group also affords an opportunity to examine the phenomenon of natural evil.

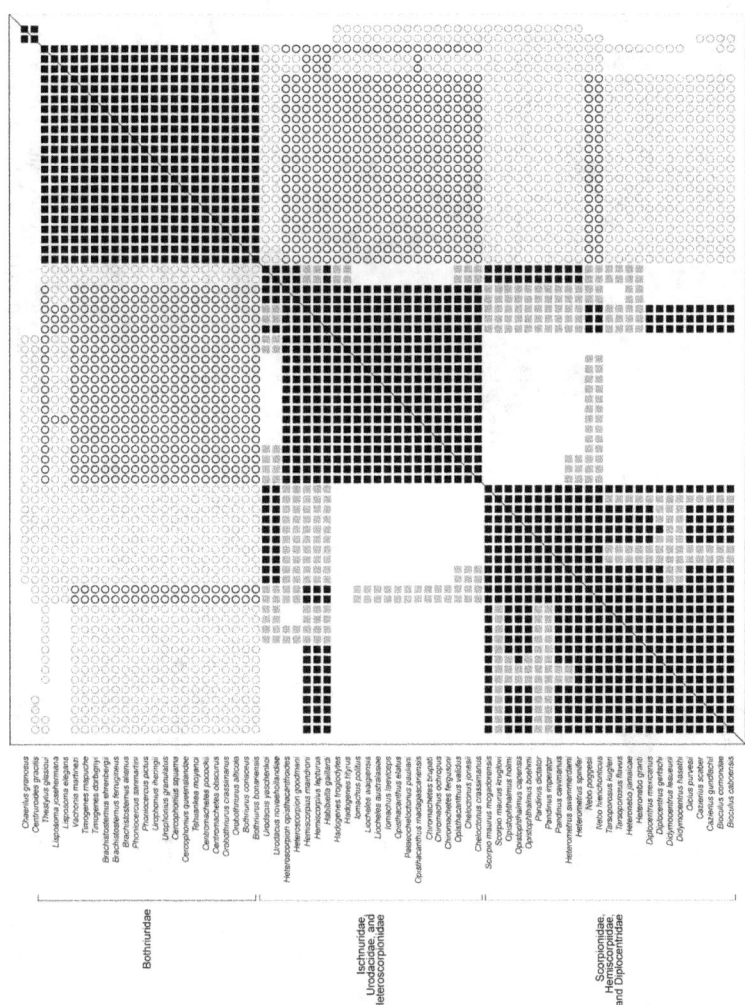

Figure 72. BDC bootstrap results for Scorpionoidea, as calculated by BDISTMDS (relevance cutoff 0.95). Closed square indicate significant, positive BDC; open circles indicate significant, negative BDC. Black symbols indicate bootstrap values >90% in a sample of 100 pseudoreplicates. Grey symbols represent bootstrap values ≤90%.

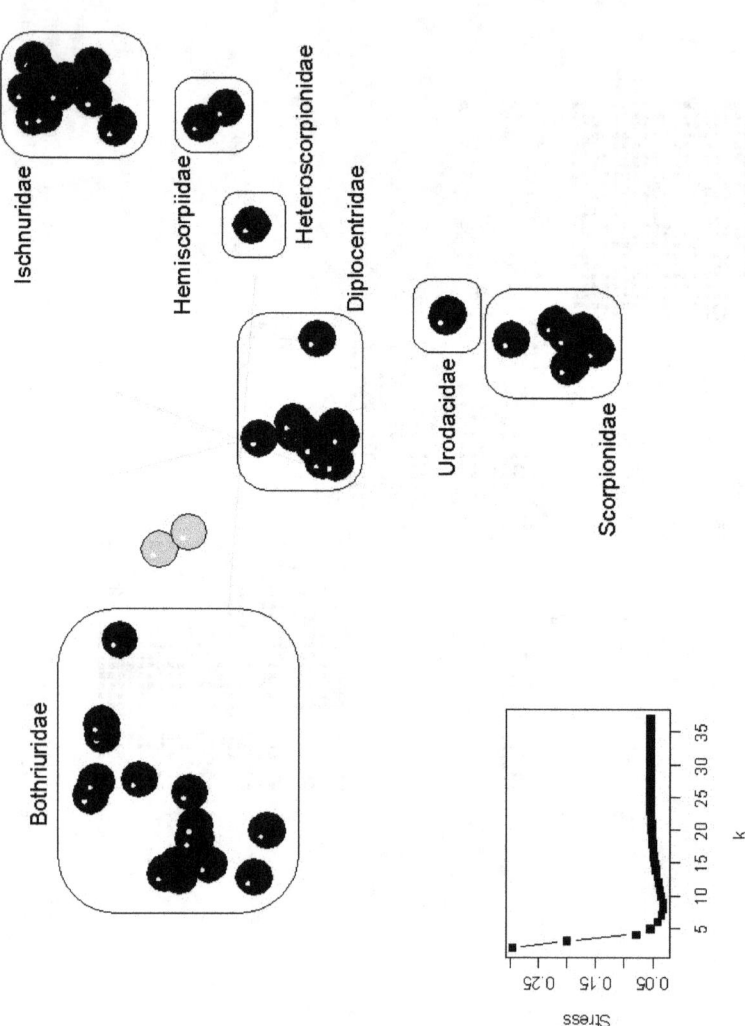

Figure 73. Three-dimensional MDS applied to Scorpionoidea baraminic distances and the stress of $k$-dimensional MDS on the same baraminic distance matrix plotted as a function of the number of dimensions ($k$). Scorpionoids are shown in black, and the outgroup taxa in gray.

## 2.35. Histeridae (Arthropoda: Insecta: Coleoptera)

Dataset published by Caterino and Vogler (2002)

| | |
|---|---|
| Characters in published dataset: | 52 |
| Taxa in published dataset: | 41 |
| Character relevance cutoff: | 0.95 |
| Characters used to calculate BD: | 35 |
| Taxa used in BDC and MDS analysis: | 41 |
| Stress for 3D MDS: | 0.207 |
| $k_{min}$: | 6 |
| Median bootstrap value: | 69 |
| $F_{90}$: | 0.246 |

Histerid beetles have diverse morphologies and occupy a diverse array of habitats. Approximately 3900 species in 330 genera are assigned to the family, including bark beetles, wood borers, inquilines, and dung and carrion eaters. This vast diversity has been classified into eleven subfamilies, the relationships of which are in some dispute (Caterino and Vogler 2002). Caterino and Vogler's (2002) morphological phylogeny was based on 52 adult and larval characters from 37 histerids and four outgroups. The histerids represented all eleven subfamilies and major tribes within those subfamilies. The morphological characters were largely exoskeletal. Their results cast doubt on previous work, although they emphasized that even their combined molecular and morphological analysis was still preliminary.

All taxa and 35 characters were used to calculate baraminic distances (omitted: all 15 larval characters and characters 49 and 51). BDC results show two main groups, one of which corresponds to the Histeridae and the other to the outgroup taxa (Figure 74). The Histeridae are a heterogeneous group with two subgroups visible. The smaller histerid subgroup consists of the subfamilies Abraeinae, Trypanaeinae, and Trypeticinae, as well as two genera of Dendrophilinae, *Bacanius* and *Anapleus*. The larger histerid subgroup contains the remaining histerid taxa. Between the subgroups, 38 taxon pairs have significant, positive BDC and eight have signficant, negative BDC. All eight taxon pairs with significant, negative BDC involve the dendrophiline *Bacanius*. Of the 148 taxon pairs involving a histerid and one of the four outgroup taxa, 101 have significant, negative BDC and one has significant, positive BDC. The outgroup *Syntelia* is positively correlated with the histerid *Onthophilus*. Bootstrap results are fairly poor for this dataset (median

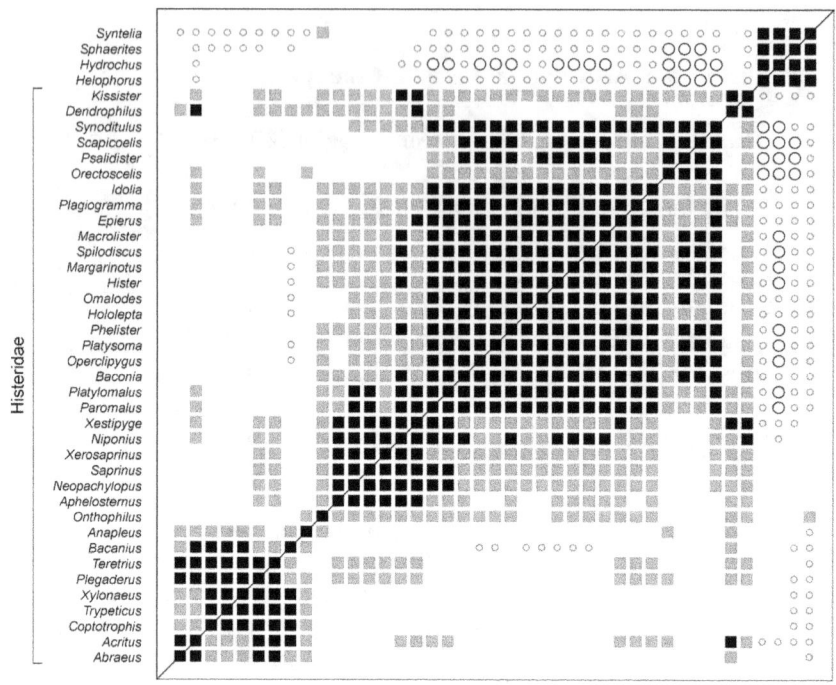

Figure 74. BDC bootstrap results for Histeridae, as calculated by BDISTMDS (relevance cutoff 0.95). Closed square indicate significant, positive BDC; open circles indicate significant, negative BDC. Black symbols indicate bootstrap values >90% in a sample of 100 pseudoreplicates. Grey symbols represent bootstrap values ≤90%.

69%), but some bootstrap values >90% occur within the subgroups of histeridae and in 20 of the outgroup/histerid taxon pairs.

The MDS results are also poor, with a 3D stress of 0.207. Nevertheless, two separated groups are immediately apparent: the Histeridae and the outgroups (Figure 75). Within the Histeridae, the subgroups detected in the BDC analysis are not apparent. Though well-separated, the nearest ingroup and outgroup taxa are *Onthophilus* and *Syntelia* respectively.

The results of these analyses seem straightforward: The Histeridae, even though sparsely sampled, appear to be a holobaramin. The positive BDC within the group supports continuity of the sampled genera (and hence the sampled subfamilies), and the negative BDC between the histerids and the outgroup taxa supports a discontinuity around Histeridae. The positive BDC between *Onthophilus* and *Syntelia* is likely to be spurious, since they are definitely parts of two different clusters in the MDS results. This is one of the largest animal baramins identified to date, although more detailed studies in the future should include additional histerid diversity to confirm the membership of the full holobaramin.

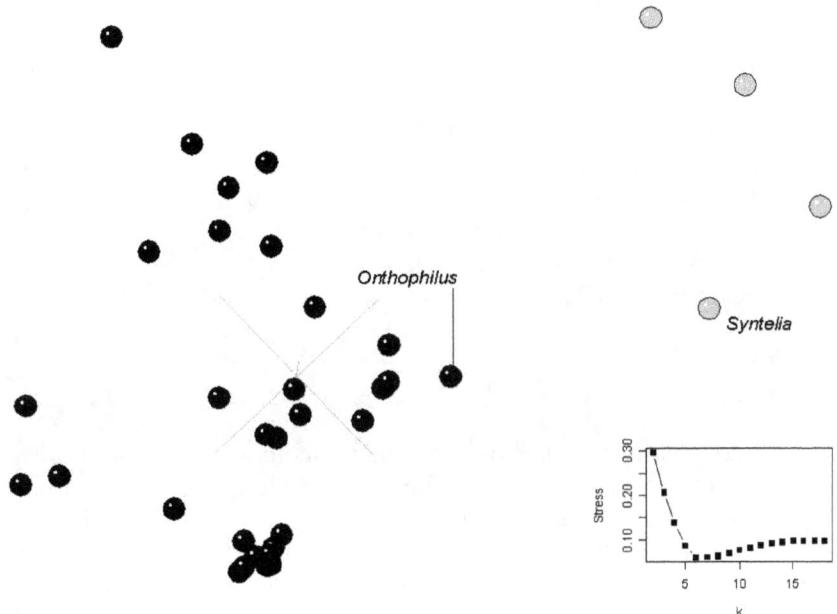

Figure 75. Three-dimensional MDS applied to Histeridae baraminic distances and the stress of k-dimensional MDS on the same baraminic distance matrix plotted as a function of the number of dimensions (k). Histerids are shown in black, outgroup taxa in gray.

## 2.36. Coelopidae (Arthropoda: Insecta: Diptera)

Dataset published by Meier and Wiegmann (2002)

| | |
|---|---|
| Characters in published dataset: | 39 |
| Taxa in published dataset: | 29 |
| Character relevance cutoff: | 0.95 |
| Characters used to calculate BD: | 39 |
| Taxa used in BDC and MDS analysis: | 29 |
| Stress for 3D MDS: | 0.116 |
| $k_{min}$: | 6 |
| Median bootstrap value: | 74 |
| $F_{90}$: | 0.352 |

In contrast to the speciose Histeridae, there are only about thirty species of Coelopidae, also called the seaweed flies. The larvae of these dipterans live on decaying seaweed, especially brown algae. Meier and Wiegmann (2002) studied the phylogeny of Coelopidae using molecular and morphological characters. Their morphological dataset consisted of 39 characters and 29 taxa. The characters were mostly exoskeletal with some soft tissue and life history attributes. The taxon sample included five dipteran outgroups and 24 coelopid species from twelve genera. Meier and Wiegmann (2002) found 84 most parsimonious trees, which supported the monophyly of Coelopidae but left several intrafamilial relationships unresolved.

All taxa and characters were used to calculate baraminic distances. Two groups appear in the BDC results (Figure 76), the Coelopidae and the outgroup taxa. Within each group, significant, positive BDC is common. The coelopids are more diverse, with only 142 of the 276 possible coelopid-only taxon pairs positively correlated. Although positive correlation is sparse, almost the entire group can be connected by a single taxon; *Icaridion debile* is positively correlated with 21 of the 23 other coelopids. No negative BDC between coelopids was observed. The outgroups are not positively correlated with any of the coelopids, and 72 of the outgroup/coelopid taxon pairs have significant, negative BDC. The bootstrap values are fairly high for the positive BDC within the coelopids (87 taxon pairs had bootstrap values >90%) and for the negative BDC between the coelopids and the outgroups (38 taxon pairs with bootstrap values >90%). The MDS results confirm the presence of two clusters of taxa (Figure 77). The coelopids appear as a triangular cloud of taxa, and all five outgroup taxa form a separate cluster.

# Animals

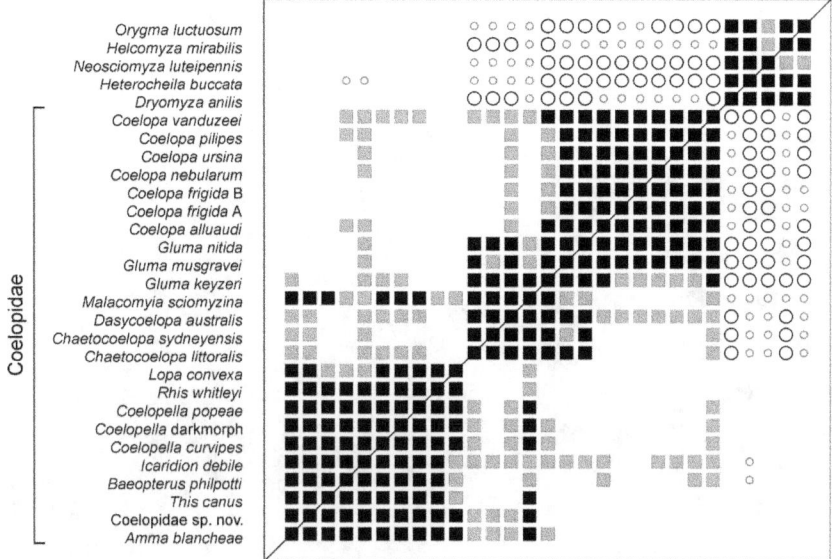

Figure 76. BDC bootstrap results for Coelopidae, as calculated by BDISTMDS (relevance cutoff 0.95). Closed square indicate significant, positive BDC; open circles indicate significant, negative BDC. Black symbols indicate bootstrap values >90% in a sample of 100 pseudoreplicates. Grey symbols represent bootstrap values ≤90%.

The positive BDC within Coelopidae supports the conclusion of continuity between the coelopid species and genera. Negative BDC with the outgroup taxa supports the conclusion of discontinuity between the Coelopidae and other dipterans. Consequently, the Coelopidae appears to be a holobaramin. Since most of the coelopid species have been sampled in this study, the only way to improve the study would be to increase the sample of characters. With the present data, however, the holobaraminic classification of the Coelopidae is well-supported.

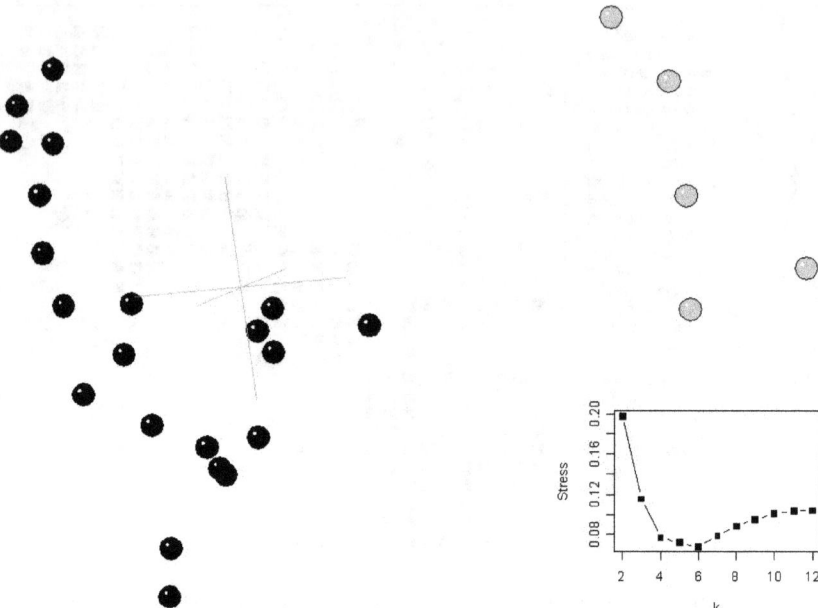

Figure 77. Three-dimensional MDS applied to Coelopidae baraminic distances and the stress of k-dimensional MDS on the same baraminic distance matrix plotted as a function of the number of dimensions (k). Coelopids are shown in black, outgroup taxa in gray.

## 2.37. Lophopidae (Arthropoda: Insecta: Hemiptera)

Dataset published by Soulier-Perkins (2001)

| | |
|---|---|
| Characters in published dataset: | 73 |
| Taxa in published dataset: | 41 |
| Character relevance cutoff: | 0.95 |
| Characters used to calculate BD: | 58 |
| Taxa used in BDC and MDS analysis: | 41 |
| Stress for 3D MDS: | 0.193 |
| $k_{min}$: | 7 |
| Median bootstrap value: | 91 |
| $F_{90}$: | 0.501 |

Lophopidae is an exclusively tropical family of treehoppers, found in southeast Asia, Africa, and South America. Soulier-Perkins (1998) lists 37 genera in Lophopidae, 32 of which are found in southeast Asia or the Pacific islands. Although several members of the family are crop pests on sugarcane or coconut palms, little is known about most of the species' plant preferences. Soulier-Perkins (2001) evaluated the phylogeny of the lophopids using a matrix of 73 characters scored for five outgroup taxa and 36 lophopids (representing 33 of the 37 genera). The characters were derived largely from external morphology as well as male and female genitalia. Soulier-Perkins found nine most parsimonious trees, none of which supported the monophyly of Lophopidae. Two South American lophopids, *Hesticus* and *Silvanana*, consistently grouped with the outgroup taxa.

For baraminic distances, all taxa and 58 of the characters were used for calculations (omitted characters: 51, 60-73). Three groups of taxa were evident in the BDC results (Figure 78). The first group consists of the outgroups *Tettigometra*, *Ricania*, and *Pochazia* together with the two "lophopids" *Silvanana* and *Hesticus*. The second group contains only the two remaining outgroups *Loxocephala* and *Aspidonitys*. The final group contains the remaining lophopids. Within the largest group, the remaining lophopids, 520 of the possible 561 taxon pairs have significant, positive BDC, and 369 of those have bootstrap values >90%. The larger group of outgroups that includes *Hesticus* and *Silvanana* are negatively correlated with the remaining lophopids, including signficiant, negative BDC between *Hesticus/Silvanana* and all other lophopids except for the Indonesian *Asantorga*. The smaller set of outgroups, *Loxocephala/Aspidonitys*, is positively correlated with only one other taxon *Asantorga* and negatively correlated with no taxa. *Asantorga* appears to be a slightly unusual lophopid, in that it is positively correlated with only 27 of the remaining 33 lophopids in the larger lophopid group, it is positively correlated with the outgroup *Aspidonitys*, and it shares no significant, negative BDC with any outgroups.

The MDS results are somewhat poor with a 3D stress of 0.193, but the distribution of taxa does support the findings of the BDC analysis (Figure 79). The Lophopidae form an elongated cloud with three outliers: *Asantorga*, *Hesticus*, and *Silvanana*. *Asantorga* is adjacent to nothing in particular, but *Hesticus* and *Silvanana* form a diffuse cluster with the two outgroups *Ricania* and *Pochazia*. The outgroup *Tettigometra* forms a diffuse cluster with the remaining outgroup taxa *Loxocephala* and *Aspidonitys*. This cluster is closer to the main lophopid cluster than the other outgroup cluster.

These results support the classification of Lophopidae *sensu stricto* (sans *Hesticus* and *Silvanana*) as a holobaramin. The BDC results alone support this, indicating that most of the species are connected by continuity (evidenced by significant, positive BDC) and separated from outgroups by discontinuity (evidenced by significant, negative BDC). The MDS results are unusual in two respects: (1) *Asantorga* is separated from the main lophopid cluster, and (2) *Tettigometra* clusters with *Loxocephala* and *Aspidonitys*. The BDC results indicate that *Asantorga* is definitely a part of the main lophopid cluster and that *Tettigometra* clusters with the *Ricania/Pochazia/Hesticus/Silvanana* group. These discrepancies are probably the result of the high stress observed for the 3D MDS. Nevertheless, the broad picture of a cluster of lophopids separated from the outgroups and from *Hesticus* and *Silvanana* is seen in both the MDS and BDC results. These results are also consistent with Soulier-Perkins's (2001) finding that *Hesticus* and *Silvanana* are not part of the monophyletic Lophopidae *sensu stricto*.

# Animals

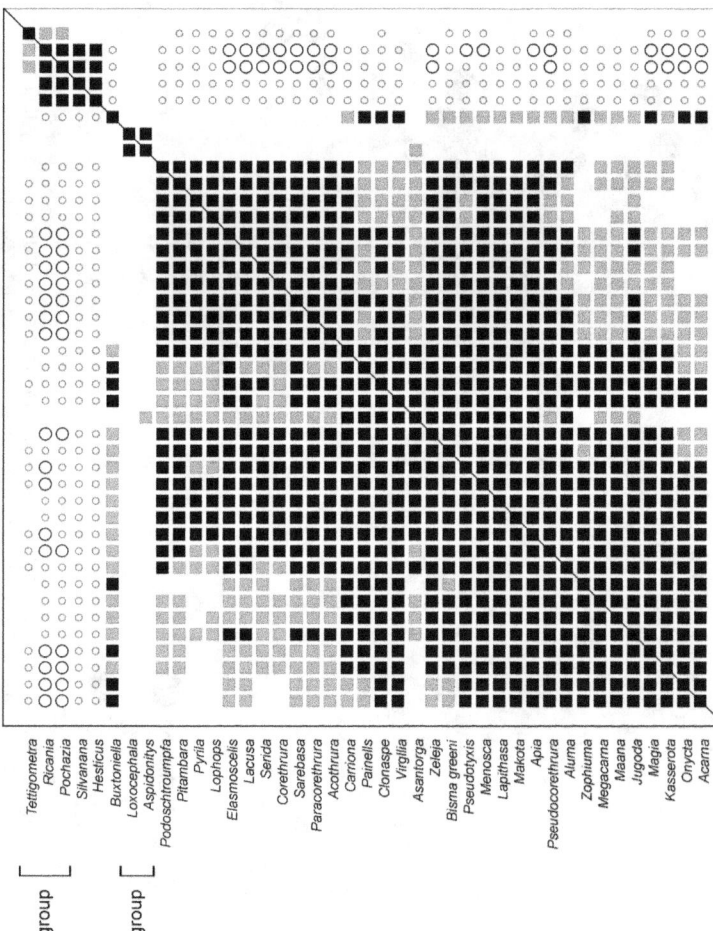

Figure 78. BDC bootstrap results for Lophopidae, as calculated by BDISTMDS (relevance cutoff 0.95). Closed square indicate significant, positive BDC; open circles indicate significant, negative BDC. Black symbols indicate bootstrap values >90% in a sample of 100 pseudoreplicates. Grey symbols represent bootstrap values ≤90%.

140   Animal and Plant Baramins

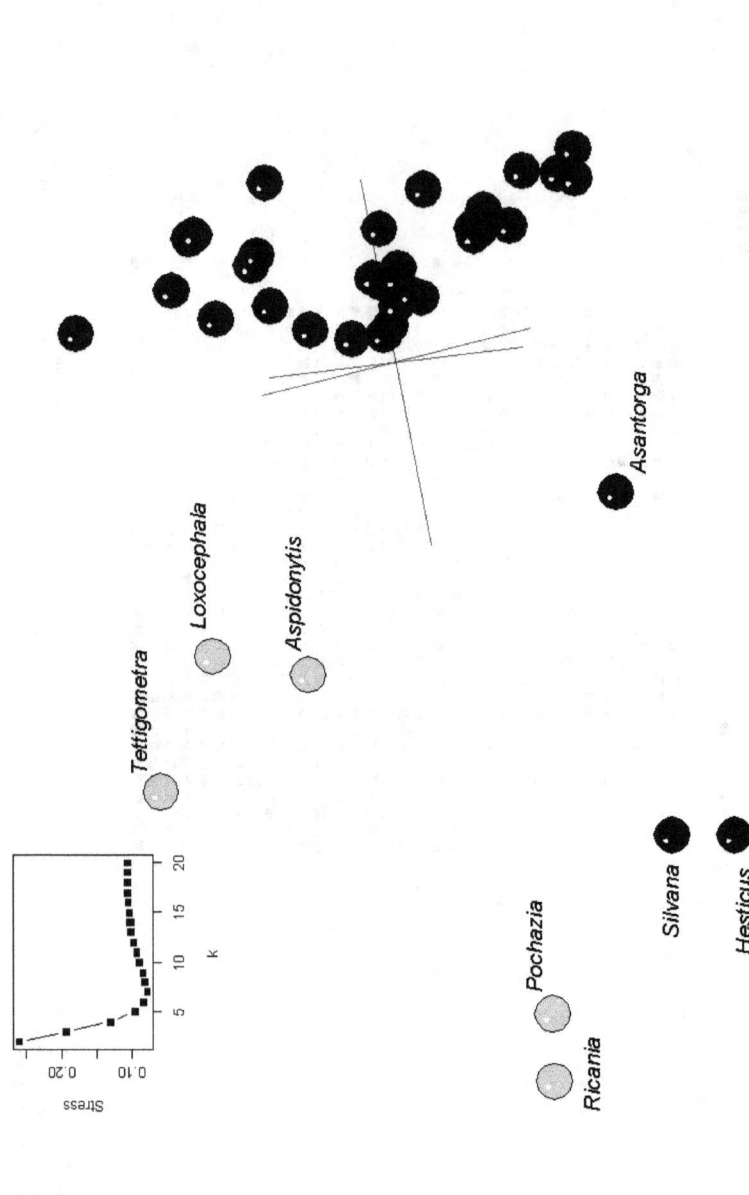

Figure 79. Three-dimensional MDS applied to Lophopidae baraminic distances and the stress of *k*-dimensional MDS on the same baraminic distance matrix plotted as a function of the number of dimensions (*k*). Lophopids are shown in black, other taxa in gray.

## 2.38. Membracidae (Arthropoda: Insecta: Hemiptera)

Dataset published by Cryan et al. (2004)

| | |
|---|---|
| Characters in published dataset: | 83 |
| Taxa in published dataset: | 66 |
| Character relevance cutoff: | 0.95 |
| Characters used to calculate BD: | 67 |
| Taxa used in BDC and MDS analysis: | 66 |
| Stress for 3D MDS: | 0.202 |
| $k_{min}$: | 8 |
| Median bootstrap value: | 63 |
| $F_{90}$: | 0.276 |

There are more than 3000 species of treehoppers and thorn bugs in the family Membracidae. Many species exhibit a striking resemblance to plant structures, while others mimic other insects. Membracids are organized into 47 tribes and nine subfamilies. Their phylogeny has been examined by molecular and morphological analyses, all of which support the monophyly of Membracidae but which differ on the phylogenetic relationships of the tribes. Cryan et al.'s (2004) morphological and molecular phylogenetic analysis of the family sampled 61 membracid species from 25 tribes and six subfamilies as well as five outgroup taxa from the families Cicadellidae and Aetalionidae. Their dataset sampled 83 morphological characters from adult and immature insects. The results of the molecular and morphological phylogenetic analysis generally agreed with respect to the monophyly of the family and the relationships of the subfamily.

To calculate baraminic distances, all taxa were used and sixteen characters were eliminated from the dataset (omitted: 19, 61-62, 70, 72-83). Two main groups are apparent in the BDC results, each with two subgroups (Figure 80). One group consists of the outgroup taxa and the membracid subfamilies Stegaspidinae, Centrodontinae (represented by one species *Centrodontus atlas*), and Centrotinae. The remaining membracid subfamilies, Darninae, Membracinae, and Smiliinae, constitute the second group. Within the first group, specific designations of subgroups is difficult because of extensive overlap. The outgroup taxa seem to form one subgroup, but two taxa *Microcentrus* and *Antillotolania* are positively correlated with every member of this group. Similarly, the stegaspidine *Lycoderes* is positively correlated with four of the outgroup taxa and seven of the other membracid taxa in this group.

The subgroups could be most easily defined by the high bootstrap values (>90%). The stegaspidines *Deiroderes* and *Antillotolania* would group with the outgroups and the remaining membracids would constitute the second subgroup. There is no significant, negative BDC between any two members of this group.

The second major group of taxa in the BDC results are composed of the membracid subfamilies Darninae, Membracinae, and Smiliinae, which form an unnamed clade in the molecular and morphological phylogenies of Cryan et al. (2004). Within this group, the two subgroups roughly correspond to the Membracinae and the Smiliinae + Darninae, but again we see significant overlap. The darnine *Cymbomorpha* is positively correlated with every member of the Membracinae, Darninae, and Smiliinae, and three taxa of the Smiliinae + Darninae (*Darnis, Micrutalis,* and *Acutalis*) have significant, positive BDC with members of the Membracinae. Bootstrap values for most of the positive BDC, even between the two subgroups, is high (>90%). Significant, negative BDC occurs in two taxon pairs: *Polyglypta/Ochropepla* and *Polyglypta/Scalmophorus*, each involving members of two different subfamilies (Smiliinae and Membracinae).

Between the two main groups of taxa in the BDC results, significant, negative BDC is common, particularly with members of the Smiliinae + Darninae. The outgroups and stegaspidines are negatively correlated with members of Membracinae, Darninae, and Smiliinae, but *Centrodontus* and the centrotines are negatively correlated only with Smiliinae and Darninae. One instance of significant, positive BDC connects the two main groups, between *Centrodontus* and the membracine *Scalmophorus*. The bootstrap value for this positive correlation was only 71%. Most of the negative BDC had bootstrap values <90%, with the exception of 23 instances, 20 of which involve the stegaspidine *Microcentrus* and members of the Smiliinae.

The MDS results are consistent with the findings of the BDC analysis, even though the MDS results are not particularly good (3D stress = 0.202). Taxa of the Membracinae, Smiliinae, and Darninae form an elongated cluster orthogonal to an elongated cluster of the other membracid subfamilies and the outgroup taxa (Figure 81). While the Membracinae, Smiliinae, and Darninae are tightly clustered, the other taxa are more diffuse. Particularly, Stegaspidinae is much more spread out than any other group of taxa. Between the two main groups, *Centrodontus* is the closest of the outgroup + membracid taxa to the taxa of Membracinae, Smiliinae, and Darninae, although it is not adjacent to Scalmophorus, with which it shares significant, positive BDC. Apart from their orthogonality, the two groups are not particularly separated.

From a baraminological standpoint, the BDC and MDS results seem to support the holobaraminic status of a single group composed of the membracid subfamilies Membracinae, Smiliinae, and Darninae (hereafter Membracidae *sensu stricto*). Within this group, continuity is evidenced by significant, positive BDC, and discontinuity with other taxa is evidenced by significant, negative BDC. MDS provides only partial support for this conclusion in that the taxa appear to form two orthogonal, linear clusters that are not noticeably separated. This pattern has been seen before in the phalacrocoracids, wherein negative BDC was not interpreted as true discontinuity but instead due to the unusual geometry (Wood 2005a). This case differs from that of the phalacrocoracids in the 3D stress. For the cormorants, 3D stress was only 0.11, but in this case, the 3D stress is nearly twice that, 0.202. Thus, the actual distribution of taxa in the 3D MDS might be a distortion of the true distribution in higher dimensions with lower stress. A distorted perspective of the taxa at three dimensions could also explain why there is significant, positive BDC between *Scalmophorus* and *Centrodontus* even though they are not adjacent in the 3D MDS results. Furthermore, the orthogonality of the phalacrocoracids was more regular than that seen in this dataset. The phalacrocoracids formed five very tightly-clustered groups, where these MDS results are more diffuse.

These geometric considerations and past experience with phalacrocoracids renders the holobaraminic status of Membracidae *sensu stricto* uncertain. This uncertainty is enhanced by the small sample of taxa in the present study. The 3000 species are here represented by only 61. The 47 tribes are represented by only 25, and the nine subfamilies are represented by only six. It is possible that future studies with a fuller sample of membracids will reveal a different picture of membracid baraminology. Given the present results, however, Membracidae *sensu stricto* can provisionally be classified as a tentative holobaramin.

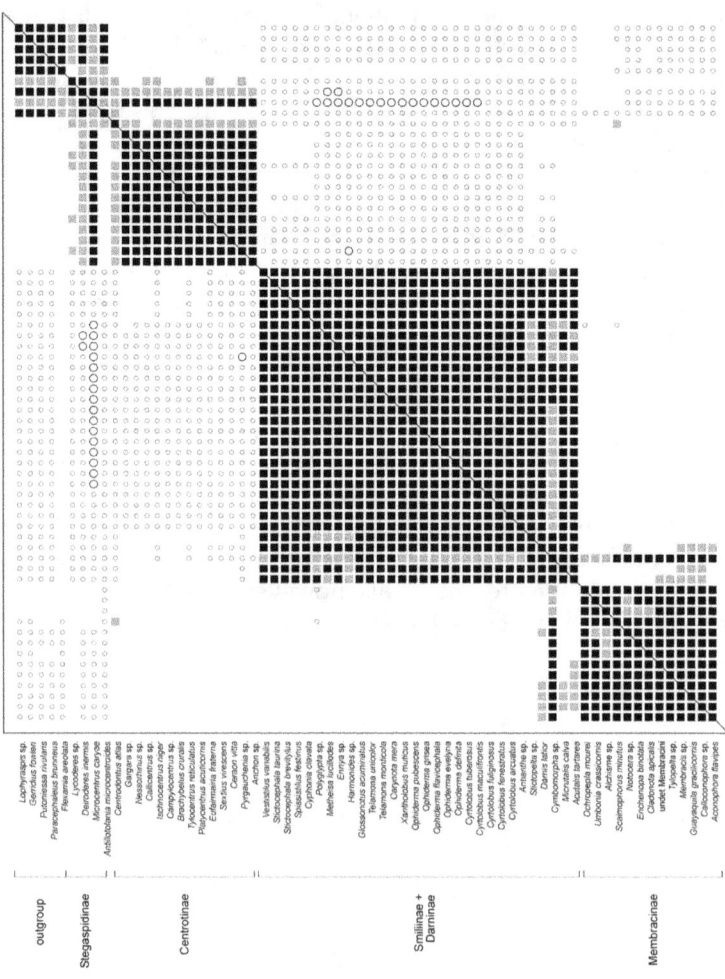

Figure 80. BDC bootstrap results for Membracidae, as calculated by BDISTMDS (relevance cutoff 0.95). Closed square indicate significant, positive BDC; open circles indicate significant, negative BDC. Black symbols indicate bootstrap values > 90% in a sample of 100 pseudoreplicates. Grey symbols represent bootstrap values ≤90%.

# Animals 145

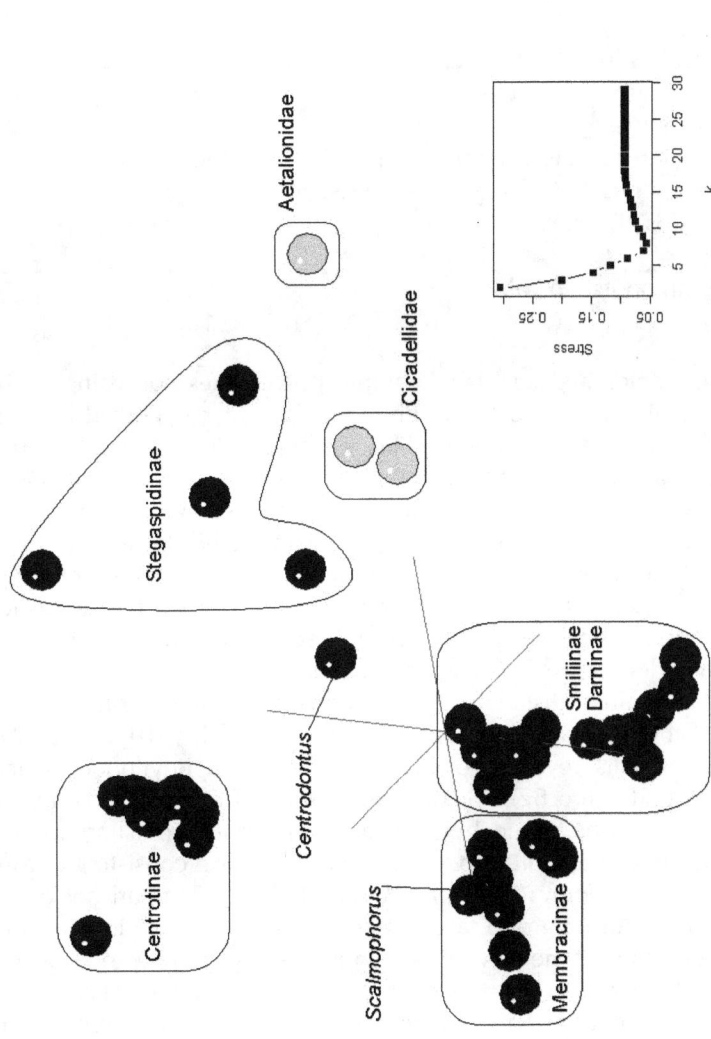

Figure 81. Three-dimensional MDS applied to Membracidae baraminic distances and the stress of $k$-dimensional MDS on the same baraminic distance matrix plotted as a function of the number of dimensions ($k$). Membracids are shown in black, outgroup taxa in gray

## 2.39. Phyllodocidae (Annelida: Polychaeta: Phyllodocida)

Dataset published by Eklöf et al. (2007)

| | |
|---|---|
| Characters in published dataset: | 29 |
| Taxa in published dataset: | 21 |
| Character relevance cutoff: | 0.95 |
| Characters used to calculate BD: | 20 |
| Taxa used in BDC and MDS analysis: | 21 |
| Stress for 3D MDS: | 0.156 |
| $k_{min}$: | 5 |
| Median bootstrap value: | 72 |
| $F_{90}$: | 0.167 |

Phyllodocids are aciculate, benthic polychaetes consisting of 18 genera and about 500 species. Phyllodocids are recognized by four anterior cirri that look something like antennae. Some pelagic groups have been classified with the phyllodocids, but according to Eklöf et al.'s (2007) molecular and morphological analysis, these groups are either basal to the benthic species or not part of the family at all. Eklöf et al.'s (2007) morphological dataset consisted of 29 external morphology characters and 21 taxa. The phyllodocid taxa sampled 19 diverse species from 13 genera. The two outgroup species represented two other families of the Phyllodocida, Nephytidae and Lacydoniidae.

For baraminic distances, calculations were made on all taxa and 20 of the 29 characterrs (omitted characters: 5, 7-10, 12-13, 21, 28). BDC results reveal two groups of taxa: the phyllodocids and the outgroups (Figure 82). Significant, positive BDC connects all the phyllodocids in one group but is sparse with poor bootstrap values. The highest bootstrap values occur in a group of taxa consisting of four genera: *Sige, Pseudomystides, Eumida,* and *Eulalia.* Comparisons of the phyllodocids with outgroup taxa result in nonsignificant BDC in most cases and significant, negative BDC otherwise. No significant, positive BDC is observed between the phyllodocids and outgroup taxa. The phyllodocids appear as an elongated cluster of taxa in the MDS results (Figure 83). The outgroup taxa form a separate cluster.

Given the small sample of taxa and characters, these results only weakly support the holobaraminic status of Phyllodocidae. Furthermore, the poor bootstrap values indicate that the results here are sensitive to character resampling and are therefore strongly dependent on the particular characters used. Nevertheless, the clustering of the

# Animals

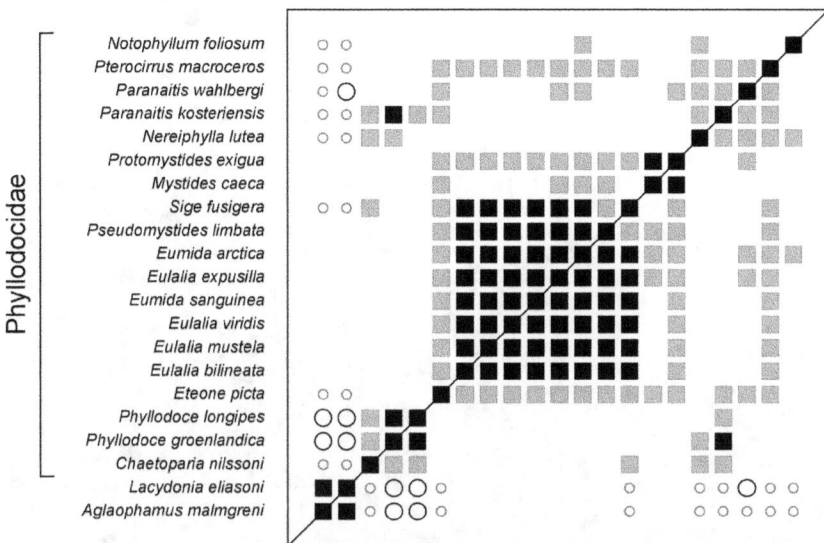

Figure 82. BDC bootstrap results for Phyllodocidae, as calculated by BDISTMDS (relevance cutoff 0.95). Closed square indicate significant, positive BDC; open circles indicate

phyllodocids in MDS and the significant, positive BDC support the classification of Phyllodocidae as a monobaramin. The significant, negative BDC between the ingroup and outgroup and the separate clustering of the outgroup in MDS support a discontinuity between the phyllodocids and other members of Phyllodocida. Taken together, these results indicate that Phyllodicidae is a holobaramin, but further research is needed to verify this conclusion.

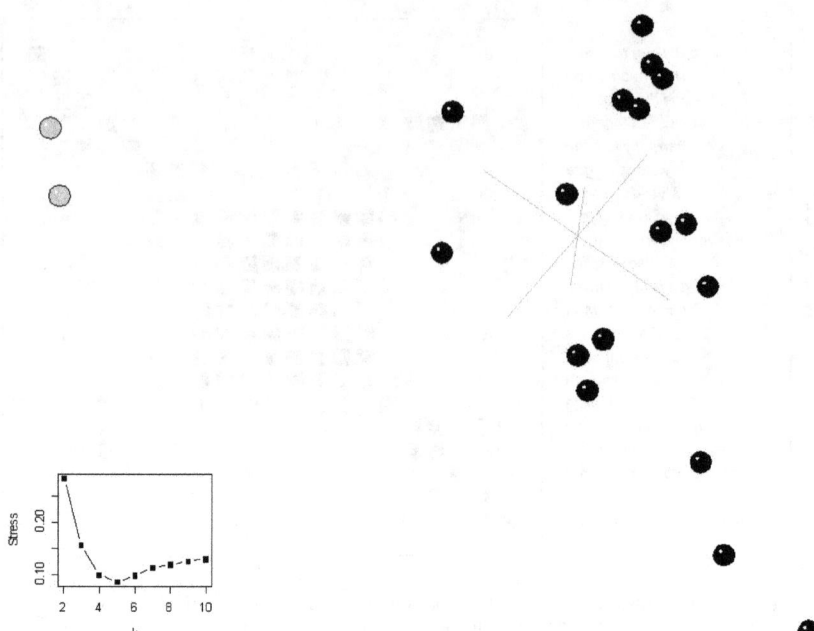

Figure 83. Three-dimensional MDS applied to Phyllodocidae baraminic distances and the stress of $k$-dimensional MDS on the same baraminic distance matrix plotted as a function of the number of dimensions ($k$). Phyllodocids are shown in black and outgroup taxa in gray.

# 3. Plantae

### 3.1. Saururaceae (Magnoliophyta: Magnoliopsida: Piperales)

Dataset published by Meng et al. (2003)

| | |
|---|---|
| Characters in published dataset: | 49 |
| Taxa in published dataset: | 7 |
| Character relevance cutoff: | 0.95 |
| Characters used to calculate BD: | 47 |
| Taxa used in BDC and MDS analysis: | 7 |
| Stress for 3D MDS: | 0.078 |
| $k_{min}$: | 3 |
| Median bootstrap value: | 99 |
| $F_{90}$: | 1 |

Saururaceae is a small family of six species in four genera. Found in North America (*Anemopsis* and *Saururus cernuus*) and Asia (*Gymnotheca, Houttuynia*, and *Saururus chinensis*), the family is sometimes called the lizard's tail family after the common name of the North American *S. cernuus*, a common flower found on the edges of lakes and streams in early summer. Meng et al. (2003) studied the phylogeny of this family using morphological and molecular characters for all six species and one outgroup (Piperaceae: *Zippelia*). Their morphological data consisted of nine vegetative and stem, 25 floral, five pollen, eight embryological, and two cytological characters, for a total of 49 characters.

For baraminic distance calculations, all seven taxa and 47 characters were used. Two vegetative and stem characters were omitted (6 and 8) due to low relevance. BDC results surprisingly reveal almost no groups at all (Figure 84). Only two taxon pairs of 21 showed any significant BDC. The two species of *Saururus* were positively correlated, as were the two species of *Gymnotheca*. Bootstrap values for both taxon pairs with positive BDC was 100%. MDS results reveal an arc of Saururaceae with *Houttuynia* at one end and *Saururus* at the other (Figure 85). The

Saururaceae
- Zippelia begoniaefolia
- Saururus chinensis
- Saururus cernuus
- Houttuynia cordata
- Gymnotheca involucrata
- Gymnotheca chinensis
- Anemopsis californica

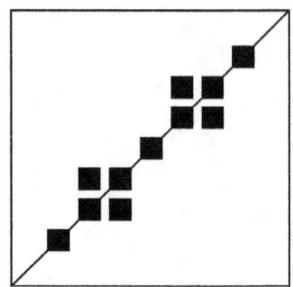

Figure 84. BDC bootstrap results for Saururaceae, as calculated by BDISTMDS (relevance cutoff 0.95). Closed square indicate significant, positive BDC..

outgroup *Zippelia* is separate and roughly equidistant from the main Saururuaceae arc.

Baraminologically, these results are noninformative. The diagnostic statistics for BDC and MDS results seem to be quite good (median bootstrap 99%; 3D stress 0.078), but the results do not reveal any significant clustering of taxa. At best, *Saururus* and *Gymnotheca* could be treated as separate monobaramins, but since they each contain only two species, that result seems somewhat trivial. Given the statistical quality of the outcome, future improvements to the baraminological research in this family are unclear. One possibility is that Saururaceae is part of a larger holobaramin that also contains Piperaceae and that additional members of Piperaceae would make the relationship more clear. Alternatively, the Saururaceae could be an unusually diverse holobaramin, and comparisons with a wider variety of outgroups could give a better picture of continuity within the family and discontinuity with other families. Finally, it is possible that the genera of Saururaceae are separate holobaramins. In such a case, additional taxa should have no impact on the results found here. Alternatively, the baraminology of Saururaceae could also be pursued by attempting artificial intergeneric hybridization to see if continuity could be established that way. Given the present data, however, the only continuity demonstrated is within the genera, and there is no conclusive evidence of discontinuity in the taxa sampled.

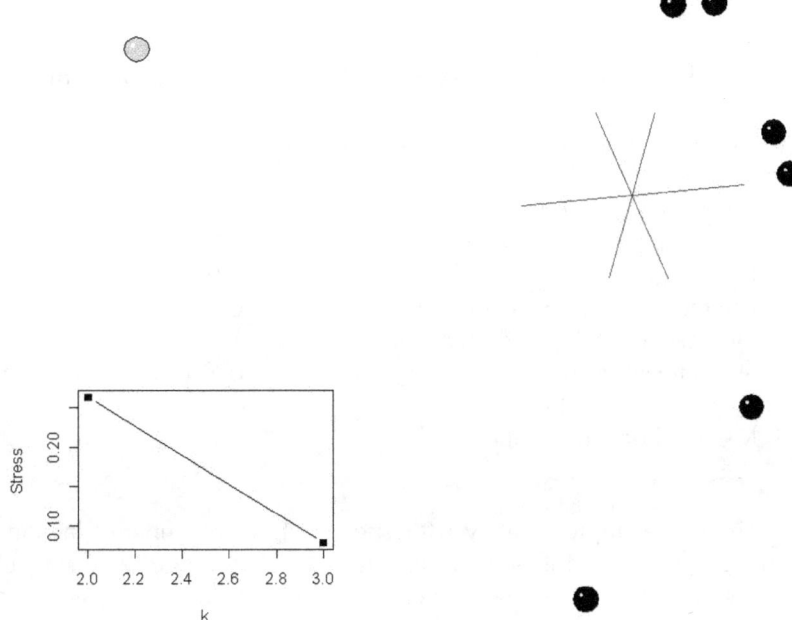

Figure 85. Three-dimensional MDS applied to Saururaceae baraminic distances and the stress of k-dimensional MDS on the same baraminic distance matrix plotted as a function of the number of dimensions (k). Saururaceae is shown in black, the outgroup in gray.

## 3.2. Aristolochiaceae (Magnoliophyta: Magnoliopsida: Aristolochiales)

Dataset published by Kelly and González (2003)

| | |
|---|---|
| Characters in published dataset: | 72 |
| Taxa in published dataset: | 18 |
| Character relevance cutoff: | 0.9 |
| Characters used to calculate BD: | 68 |
| Taxa used in BDC and MDS analysis: | 18 |
| Stress for 3D MDS: | 0.148 |
| $k_{min}$: | 7 |
| Median bootstrap value: | 86 |
| $F_{90}$: | 0.431 |

There are approximately 400 species in eight genera classified in the birthwort or dutchman's pipe family. The majority of species are placed in one genus *Aristolochia*, and the family is distinguished by its large and fetid flowers. The species are pantropical with a few temperate members, and they produce a toxic chemical that is a known carcinogen in humans and rats. Despite the apparent morphological unity of the family, molecular studies have questioned the monophyly of Aristolochiaceae (see Soltis et al. 2005, p. 59). Kelly and González (2003) evaluated the phylogeny of Aristolochiaceae and closely allied taxa using 72 morphological characters. Their taxon sample consisted of nine members of Aristolochiaceae and nine outgroups from nine different families in the Piperales, Magnoliales, Laurales, Liliales, and Arales. The characters comprised vegetative and floral attributes, life history features, and biochemical characters. Their results supported the monophyly of the Aristolochiaceae with seven synapomorphies.

All taxa and 68 characters were used to calculate baraminic distances (omitted characters: 6, 18, 44, 61). Four groups are evident in the BDC results (Figure 86). All Aristolochiaceae taxa comprise one group, with significant, positive BDC between most taxon pairs. The remaining three groups are composed of outgroup taxa. One of these groups contains both monocot taxa (*Trichopus* and *Acorus*), one contains the Piperales taxa (*Saururus*, *Piper*, and *Chloranthus*), and the remaining Magnoliales and Laurales taxa comprise the third group of outgroup taxa (*Lactoris*, *Drimys*, *Calycanthus*, and *Anaxagorea*). Between the outgroups and Aristolochiaceae, no positive BDC is observed, and significant, negative BDC occurs with the Piperales taxa and with *Acorus*.

The group containing Magnoliales and Laurales species is not positively or negatively correlated with any other taxa.

Within the Aristolochiaceae, >90% bootstrap values appear to define two separate subgroups. One subgroup consists of three species of the genera *Saruma* and *Asarum*, and the other subgroup contains the rest of the Aristolochiaceae. These subgroups correspond to the two major clades in Aristolochiaceae recovered by Kelly and González's (2004) phylogenetic study. Significant, positive BDC unequivocally connects the two subgroups in ten different taxon pairs, but the bootstrap values for those pairs is 50-73%, indicating that those connections are sensitive to random fluctuations in the characters and are therefore highly dependent on the particular characters used in this analysis.

MDS reveals two tight clusters of Aristolochiaceae, corresponding to the two subgroups found in the BDC results, adjacent to a diffuse cloud of taxa representing the outgroups (Figure 87). The groups of outgroup taxa found in the BDC results can be detected in the MDS results, but the individual genera are widely-spaced, in contrast to the Aristolochiaceae. The outgroup and Aristolochiaceae are definitely separate groups of taxa in the MDS results.

The present results support the holobaraminic classification of Aristolochiaceae. The significant, positive BDC within the group indicates continuity, and the significant, negative BDC with outgroup taxa indicates discontinuity. There are several concerns regarding this classification that will need to be addressed in future baraminology studies. First, the two subgroups of Aristolochiaceae, though connected by significant, positive BDC, appear as two separate clusters in the MDS results. A fuller sample of Aristolochiaceae taxa and/or characters should help to confirm that these two groups belong to the same holobaramin. Second, the position of the Magnoliales/Laurales outgroup taxa in this study is questionable. Although they are definitely not part of the Aristolochiaceae cluster in MDS, they do not share significant, negative BDC, indicating discontinuity. Any future studies with broader sampling of Aristolochiaceae should also narrow the focus on these outgroup taxa and omit the monocot and Piperales. Given the present data, however, Aristolochiaceae is a holobaramin.

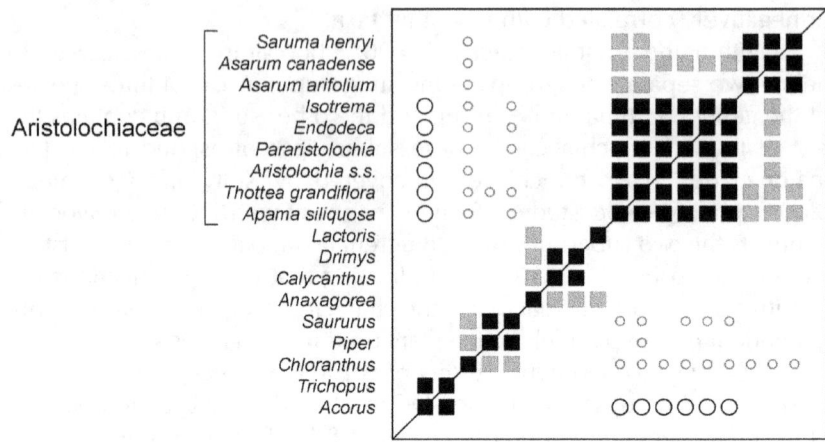

Figure 86. BDC bootstrap results for Aristolochiaceae, as calculated by BDISTMDS (relevance cutoff 0.9). Closed square indicate significant, positive BDC; open circles indicate significant, negative BDC. Black symbols indicate bootstrap values >90% in a sample of 100 pseudoreplicates. Grey symbols represent bootstrap values ≤90%.

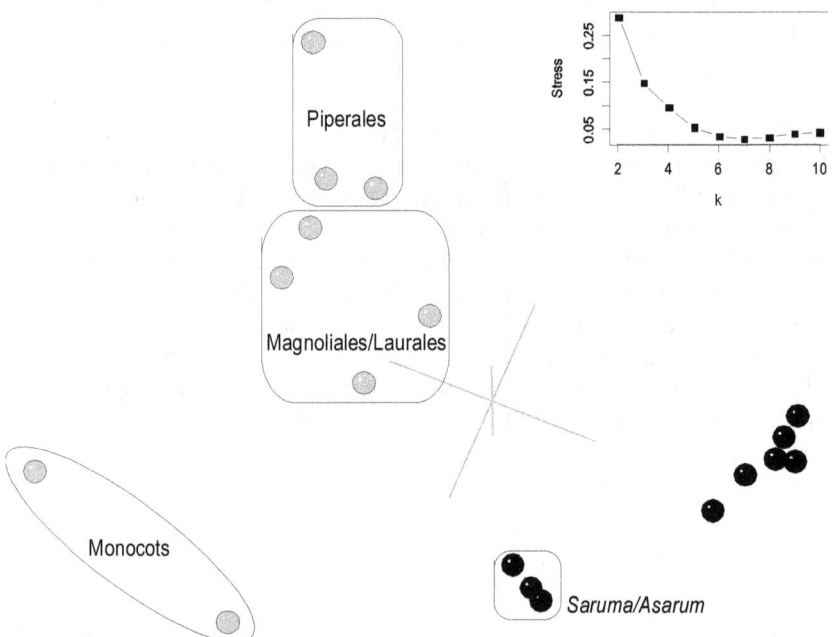

Figure 87. Three-dimensional MDS applied to Aristolochiaceae baraminic distances and the stress of $k$-dimensional MDS on the same baraminic distance matrix plotted as a function of the number of dimensions ($k$). Aristolochiaceae is shown in black, outgroup taxa in gray.

## 3.3. Nymphaeaceae (Magniolophyta: Magniolopsida: Nymphaeales)

Dataset published by Les et al. (1999)

| | |
|---|---|
| Characters in published dataset: | 68 |
| Taxa in published dataset: | 8 |
| Character relevance cutoff: | 0.95 |
| Characters used to calculate BD: | 65 |
| Taxa used in BDC and MDS analysis: | 8 |
| Stress for 3D MDS: | 0.087 |
| $k_{min}$: | 5 |
| Median bootstrap value: | 92 |
| $F_{90}$: | 0.571 |

Water lilies (family Nymphaeaceae) are aquatic perennial herbs. Included in this family are the familiar lily pads. The family consists of 60-80 species in six genera and is a member of the order Nymphaeales, which also includes Ceratophyllaceae, Cabombaceae, and Nelumbonaceae (Cronquist 1981, pp. 105-106). Les et al. (1999) evaluated the phylogeny of Nymphaeales using both molecular and nonmolecular data. Their nonmolecular data consisted of 31 vegetative/life history characters and 37 reproductive characters. All six Nymphaeaceae genera were included in the taxa sample, along with two genera of Cabombaceae as the outgroup. Their findings supported the monophyly of the Nymphaeaceae.

All taxa and 65 characters were used to calculate baraminic distances (omitted characters: 17-19). The BDC results reveal two groups of taxa: the outgroup Cabombaceae genera and the Nymphaeaceae genera *Ondinea*, *Nymphaea*, *Victoria*, and *Euryale* (Figure 88). Two Nymphaeaceae genera, *Barclaya* and *Nuphar* show neither positive nor negative BDC with any other taxa in the dataset. Cabombaceae and the group of four Nymphaeaceae genera are negatively correlated. Only two taxon pairs with positive BDC, *Victoria*/*Euryale* and *Cabomba*/*Brasenia*, have bootstrap values >90%, while four of the six taxon pairs with negative BDC have bootstrap values >90%. The negative BDC is more robust to perturbations in the data then the positive BDC. The MDS results show a diffuse cloud of taxa with two loose but separate clusters. The two clusters correspond to the Cabombaceae and Nymphaeaceae (Figure 89).

These results suggest that at least four genera of Nymphaeaceae form a holobaramin, distinct from Cabombaceae. *Nymphaea*, *Victoria*,

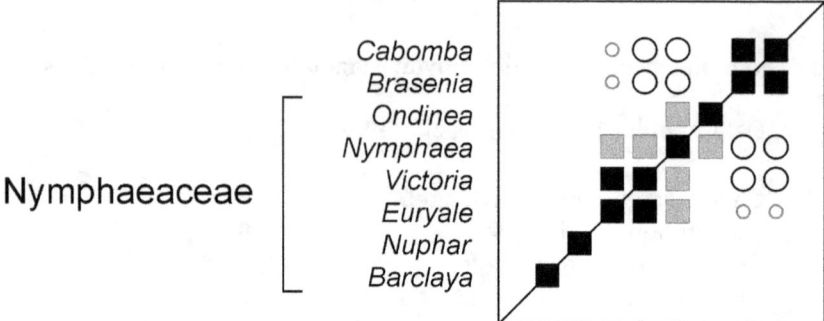

Figure 88. BDC bootstrap results for Nymphaeaceae, as calculated by BDISTMDS (relevance cutoff 0.95). Closed square indicate significant, positive BDC; open circles indicate significant, negative BDC. Black symbols indicate bootstrap values >90% in a sample of 100 pseudoreplicates. Grey symbols represent bootstrap values ≤90%.

*Euryale*, and *Ondinea* form a group connected by significant, positive BDC, suggesting continuity between the genera, and are separated from the genera of Cabombaceae by significant, negative BDC, suggesting discontinuity. The position of the two other Nymphaeaceae genera, *Barclaya* and *Nuphar* is more problematic. BDC results do not show any positive or negative correlation with any other taxa in this sample, but the MDS results suggest that they are more a part of the Nymphaeaceae cluster than they are a separate cluster or part of the Cabombaceae. Cronquist (1981, pp. 109-112) splits these genera into two different families, the monogeneric Barclayaceae (*Barclaya*) and Nymphaeaceae (*Nuphar*), but Les et al.'s (1999) phylogenetic analysis places *Nuphar* basal to *Barclaya*, and both are basal to the remaining four genera. Thus, from a phylogenetic perspective, it would seem that if *Nuphar* is part of Nymphaeaceae, *Barclaya* must be as well. Baraminology, however, is not phylogenetic, so the membership status of *Barclaya* and *Nuphar* remain unresolved. Even more provocative is the possibility that all eight genera in this dataset belong to a single holobaramin, since historically, Cabombaceae has been included as a subfamily within the Nymphaeaceae. Perhaps these relationships can be clarified by including additional outgroup taxa in future studies. At present, though, Nymphaeaceae can be tentatively classified as a holobaramin, with judgment reserved on *Barclaya* and *Nuphar*.

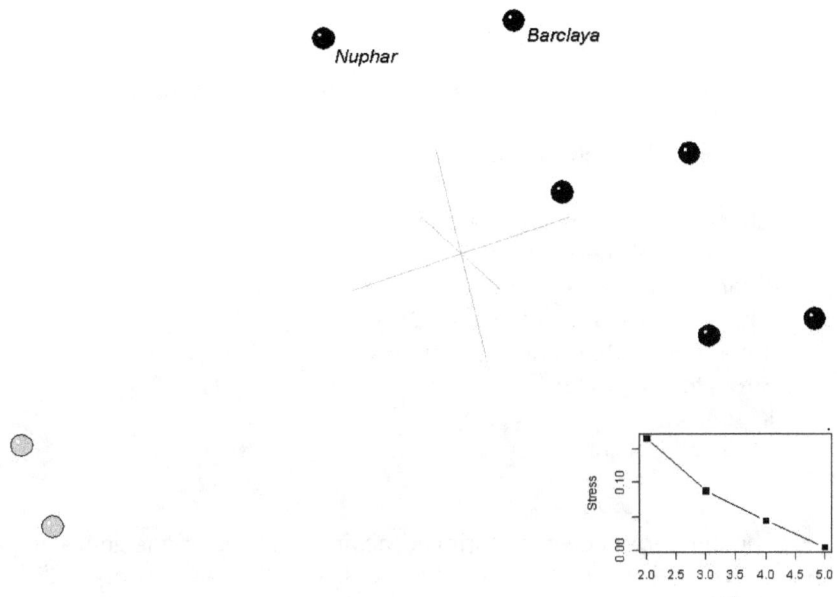

Figure 89. Three-dimensional MDS applied to Nymphaeaceae baraminic distances and the stress of $k$-dimensional MDS on the same baraminic distance matrix plotted as a function of the number of dimensions ($k$). Nymphaeaceae is shown in black, outgroup taxa in gray.

### 3.4. Moringaceae (Magniolophyta: Magniolopsida: Capparidales)

Dataset published by Olson (2002)

| | |
|---|---|
| Characters in published dataset: | 28 |
| Taxa in published dataset: | 14 |
| Character relevance cutoff: | 0.9 |
| Characters used to calculate BD: | 26 |
| Taxa used in BDC and MDS analysis: | 14 |
| Stress for 3D MDS: | 0.112 |
| $k_{min}$: | 5 |
| Median bootstrap value: | 76 |
| $F_{90}$: | 0.132 |

The thirteen species of Moringaceae are classified in one genus and are found in Africa and southwest Asia. Called bottle trees or horseradish trees, *Moringa* species are extremely variable in flower morphology and growth habit. For example, *M. drouhardii* is a tree with a bloated trunk and radially-symmetric flowers, while *M. arborea* has a slender trunk with a bilateral flower. Olson evaluated the phylogeny of *Moringaceae* using molecular and morphological data from all thirteen species and the outgroup *Cylicomorpha* of the papaya family Caricaceae. His 28 characters came from wood, bark, and roots (12), seed, seedling, and leaves (9), and flowers (7). His results showed that the bottle trees were the most basal of the *Moringa* species.

To calculate baraminic distances, all taxa and 26 characters were used (omitted characters: 16 and 26). Three groups of taxa are apparent from the BDC results (Figure 90). The first group consists of the outgroup *Cylicomorpha* and the four speices of bottletrees in section *Donaldsonia*. The second group consists of the six species of tuberous moringas, and the third group consists of three species of slender tree moringas. Within each group, there is significant, positive BDC. Between the groups, almost all taxa of the *Donaldsonia/Cylicomorpha* group and the tuberous *Moringa* group are negatively correlated. In general, the slender *Moringa* group is not correlated with taxa of any other group, except for *Cylicomorpha* and *M. drouhardii*, which are negatively correlated with *M. oleifera*. Bootstrap values are exceptionally low for most correlations (median 76%), indicating that the correlations observed depend greatly on the specific characters used in calculating baraminic distances.

The taxa appear widely-spaced with very little clustering in the MDS results (Figure 91). The bottletree moringas are found on one end

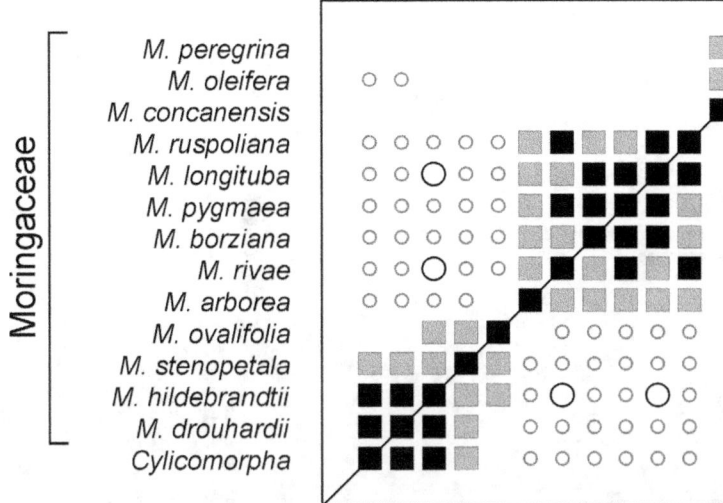

Figure 90. BDC bootstrap results for Moringaceae, as calculated by BDISTMDS (relevance cutoff 0.9). Closed square indicate significant, positive BDC; open circles indicate significant, negative BDC. Black symbols indicate bootstrap values >90% in a sample of 100 pseudoreplicates. Grey symbols represent bootstrap values ≤90%.

of the arc, with the tuberous moringas on the other and the slender tree moringas in between. None of the groups are obvious clusters separated from the main arc of taxa. The bottletree moringas (*Donaldsonia/ Cylicomorpha*) are especially diffuse in their distribution.

From a baraminological perspective, the present evidence is inconclusive. Given the MDS results only, there is no evidence of the kind of clustering usually seen in a definitive baraminology study. Therefore, it appears as though all the taxa in the sample (*Moringa* and *Cylicomorpha*) are members of a single monobaramin and discontinuity is not present. On the other hand, the BDC results would support discontinuity between the *Donaldsonia/Cylicomorpha* group and the tuberous moringas. There is also no evidence from the BDC results that connects the slender tree moringas with any of the other Moringaceae. Given this conflict between analytical techniques, it is impossible to make a conclusive identification of the Moringaceae baramin(s).

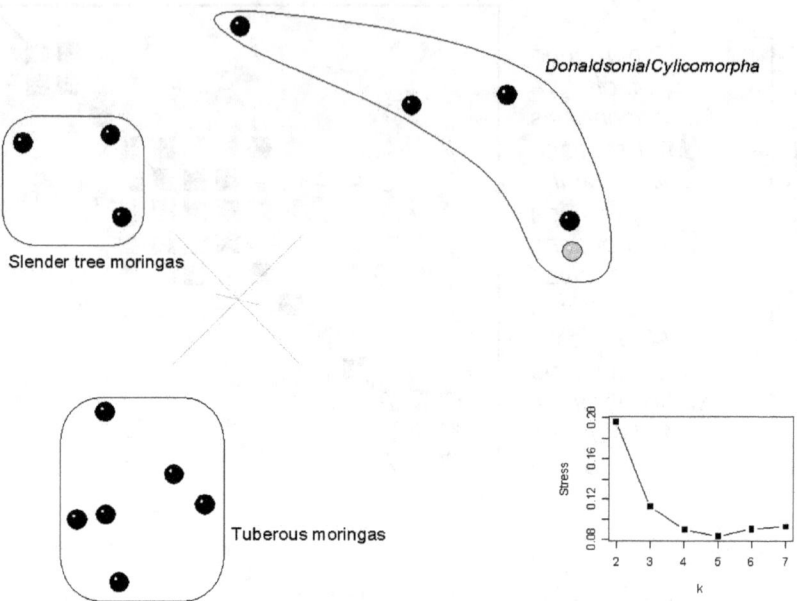

Figure 91. Three-dimensional MDS applied to Moringaceae baraminic distances and the stress of *k*-dimensional MDS on the same baraminic distance matrix plotted as a function of the number of dimensions (*k*). Moringaceae taxa appear in black and the outgroup in gray.

## 3.5. Alseuosmiaceae (Magniolophyta: Magniolopsida: Rosales)

Dataset published by Kårehed et al. (1999)

| | |
|---|---|
| Characters in published dataset: | 35 |
| Taxa in published dataset: | 17 |
| Character relevance cutoff: | 0.9 |
| Characters used to calculate BD: | 19 |
| Taxa used in BDC and MDS analysis: | 16 |
| Stress for 3D MDS: | 0.11 |
| $k_{min}$: | 4 |
| Median bootstrap value: | 72 |
| $F_{90}$: | 0.158 |

Alseuosmiaceae is a small family of Australasian shrubs consisting of approximately nine species in four genera. According to Kårehed et al.'s (1999) phylogenetic analysis, Alseuosmiaceae is a sister taxon to a clade composed of Phellinaceae and Argophyllaceae. Each of these three families are monophyletic. Kårehed et al. (1999) conducted their cladistic analysis on a matrix of 35 characters and 17 taxa, which included four species of Alseuosmiaceae from three genera. The characters consisted of leaf, trichome, flower, pollen, fruit, and wood anatomy characters. Kårehed et al. (1999) included two families in their ingroup along with Alseuosmiaceae: Phellinaceae (three taxa) and Argophyllaceae (six taxa). The remaining three taxa came from Carpodetaceae.

For baraminic distances, the taxon *Argophyllum* sp. was omitted because it had a taxic relevance of only 0.514. The remaining taxa were used to calculate baraminic distances at a character relevance cutoff of 0.9, leaving sixteen of the original 35 characters. The BDC results divided the taxa into groups based on families, with some significant, positive BDC between Carpodetaceae and Argophyllaceae (Figure 92). The only significant, negative BDC occurs between members of the Alseuosmiaceae and all other taxa. Bootstrap values are low (median 72%), indicating that the present BDC results are dependent on the particular characters chosen.

MDS results show four clusters of taxa that correspond to the four groups seen in the BDC results (Figure 93). Three of the groups form a larger triangular cluster, while Alseuosmiaceae is distinct from the other three.

The BDC and MDS results both support the holobaraminic status of Alseuosmiaceae. The significant, positive BDC within the group

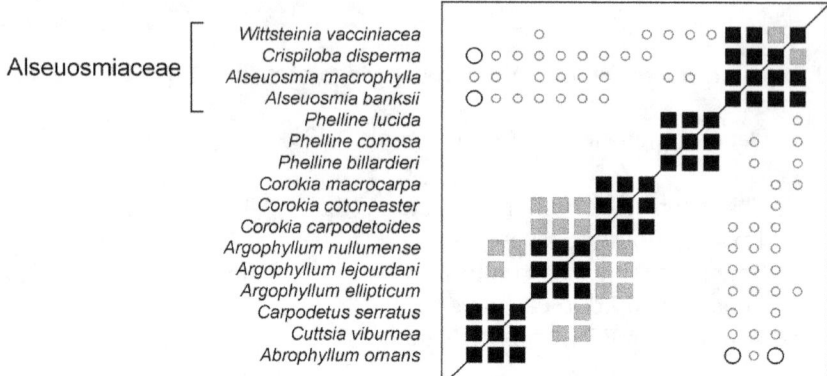

Figure 92. BDC bootstrap results for Alseuosmiaceae, as calculated by BDISTMDS (relevance cutoff 0.9). Closed square indicate significant, positive BDC; open circles indicate significant, negative BDC. Black symbols indicate bootstrap values >90% in a sample of 100 pseudoreplicates. Grey symbols represent bootstrap values ≤90%.

indicates continuity with the species, and the significant, negative BDC supports discontinuity with taxa from other families. The clustering pattern in MDS also indicates that the species of Alseosmiaceae form a cluster separated from other taxa. These results could be improved substantially by further study with the present dataset to reduce the number of unknown character states and thereby increase the number of characters. The sixteen characters used in this analysis can hardly be called a holistic sample of attributes. Future studies should also integrate the remaining species and genera from Alseuosmiaceae that were omitted from this study. Nevertheless, given the present data, Alseuosmiaceae appears to be a holobaramin.

Plants 163

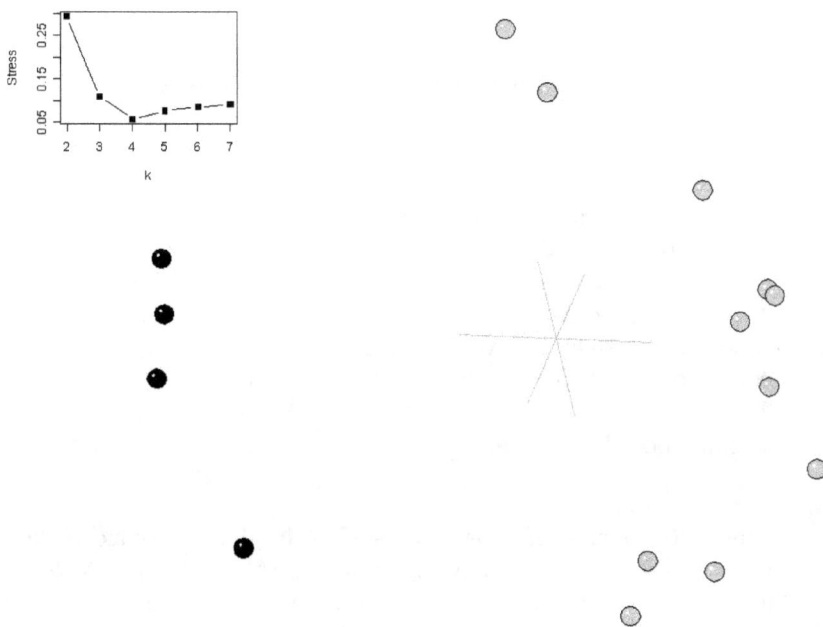

Figure 93. Three-dimensional MDS applied to Alseuosmiaceae baraminic distances and the stress of *k*-dimensional MDS on the same baraminic distance matrix plotted as a function of the number of dimensions (*k*). Alseuosmiaceae is shown in black, outgroup taxa in gray.

## 3.6. Cunoniaceae (Magniolophyta: Magniolopsida: Rosales)

Dataset published by Bradford and Barnes (2001)

| | |
|---|---|
| Characters in published dataset: | 48 |
| Taxa in published dataset: | 39 |
| Character relevance cutoff: | 0.95 |
| Characters used to calculate BD: | 44 |
| Taxa used in BDC and MDS analysis: | 38 |
| Stress for 3D MDS: | 0.281 |
| $k_{min}$: | 8 |
| Median bootstrap value: | 69 |
| $F_{90}$: | 0.148 |

The 300 species of trees and shrubs in the Cunoniaceae are classified into 26 genera. They are found mostly in temperate areas of the southern hemisphere, and their fossil record extends to the Cretaceous (Schönenberger et al. 2001). In their phylogenetic analysis, Bradford and Barnes (2001) sampled 32 Cunoniaceae taxa and seven outgroups from Cephalotaceae, Brunelliaceae, Elaeocarpaceae, and Tremandraceae. They found that the Cunoniaceae were monophyletic, and they recognized six monophyletic tribes within the Cunoniaceae. Their dataset consisted of molecular and morphological characters. They used 17 leaf and stem, 27 reproductive, and four wood anatomy characters for their morphological dataset.

Prior to baraminic distance calculation, one taxon (*Cunonia* 1) was removed because of a typographical error in the published dataset. The remaining taxa were used to calculate baraminic distances at a character relevance cutoff of 0.95, which omitted four characters (23, 30, 44, and 47). In the BDC results there appears to be one main group of taxa, which corresponds to Bradford and Barnes's (2001) tribes Cunonieae, Codieae, and Caldcluvieae, and several unclassified Cunoniaceae genera (Figure 94). A second smaller group consisting of tribes Geissoieae and Schizomerieae, together with more unclassified Cunoniaceae genera, overlaps with the larger group by significant, positive BDC. The remaining Cunoniaceae are more or less connected with these two groups by significant, positive BDC, with the exception of *Bauera*, which positively correlates with three of the outgroups and negative correlates with five Cunoniaceae species. The outgroup taxon *Brunellia* is positively correlated with *Spiraeanthemum* and *Acsmithia* (Cunoniaceae) and the outgroup *Cephalotus*. *Brunellia* is negatively correlated with *Bauera*. In addition to *Brunellia, Cephalotus* is positively correlated with four other outgroup taxa in the Tremandraceae and Elaeocarpaceae and negatively correlated with 18 taxa of Cunoniaceae. The remaining outgroup taxa (Elaeocarpaceae and Tremandraceae) form a single group that is negatively correlated with most of the Cunoniaceae except *Bauera*. Bootstrap results are poor (median 69%), indicating that much of the positive and negative BDC is sensitive to character resampling.

The Cunoniaceae appear as a diffuse cloud of taxa in the MDS results, with *Bauera* as an outlier (Figure 95). The outgroup *Brunellia* can be found well within the main cluster of Cunoniaceae. The remaining outgroup taxa form a diffuse cloud around one side of the Cunoniaceae. Although these clustering patterns are consistent with the BDC results, the 3D MDS results have a stress of 0.28, indicating that they are not the best representation of the distribution of taxa in character space.

Apart, the BDC and MDS results are somewhat difficult to interpret, but taken together, they seem to indicate that a group composed of *Brunellia* and the Cunoniaceae except *Bauera* is a holobaramin. The MDS results easily support this classification, but the high 3D stress indicates that taxic distribution could be slightly different in higher dimensional MDS. The BDC results are somewhat ambiguous, with *Brunellia* being positively correlated with two Cunoniaceae and the outgroup *Cephalotus*. Based on the MDS results, however, the significant, positive BDC between *Brunellia* and *Cephalotus* is probably spurious, since they are not adjacent and *Brunellia* is part of the main Cunoniaceae cluster. The removal of *Bauera* from the Cunoniaceae is supported in both the MDS and BDC results. Given the continuity of Cunoniaceae species and *Brunellia* as evidenced by significant, positive BDC and the discontinuity with the outgroup and *Bauera* as evidenced by significant, negative BDC, Cunoniaceae (including *Brunellia* excluding *Bauera*) is a holobaramin.

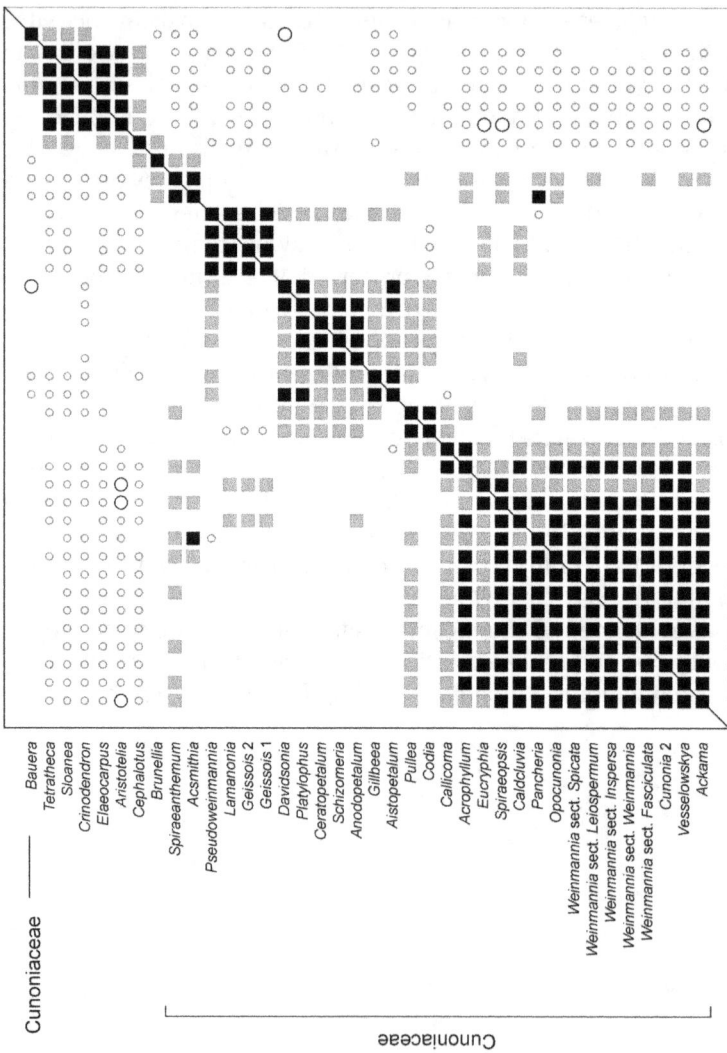

Figure 94. BDC bootstrap results for Cunoniaceae, as calculated by BDISTMDS (relevance cutoff 0.95). Closed square indicate significant, positive BDC; open circles indicate significant, negative BDC. Black symbols indicate bootstrap values >90% in a sample of 100 pseudoreplicates. Grey symbols represent bootstrap values ≤90%.

# Plants 167

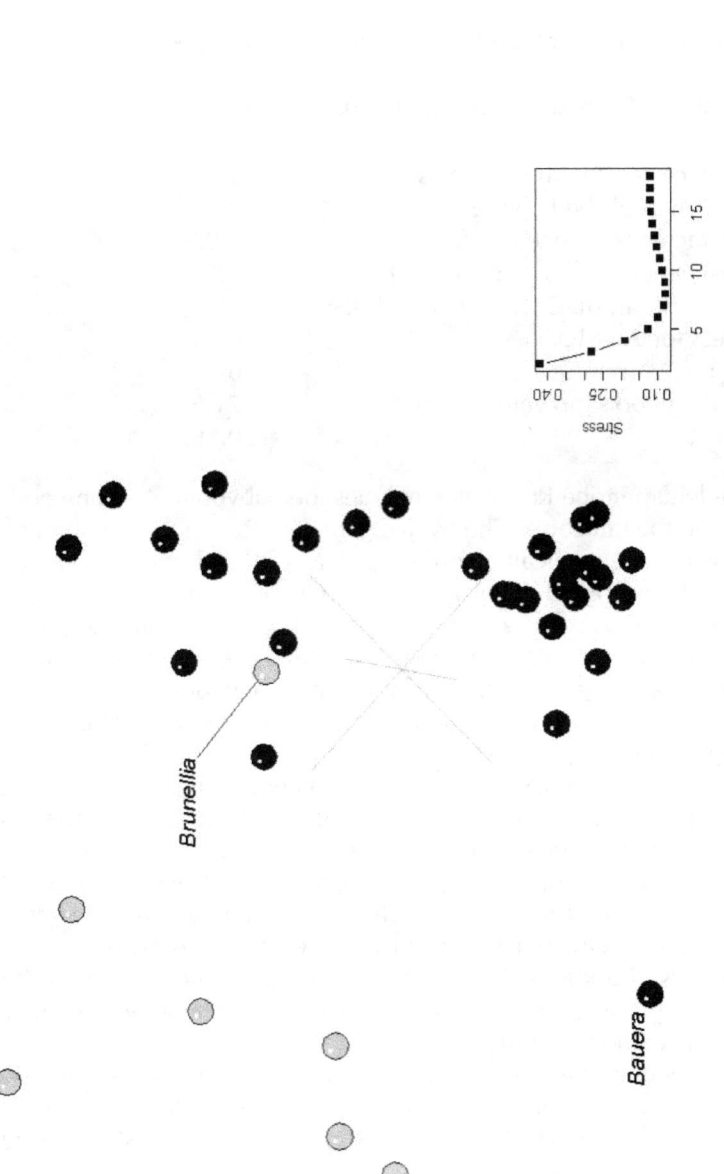

Figure 95. Three-dimensional MDS applied to Cunoniaceae baraminic distances and the stress of $k$-dimensional MDS on the same baraminic distance matrix plotted as a function of the number of dimensions ($k$). Cunoniaceae is shown in black, outgroup taxa in gray.

### 3.7. Robinieae (Magniolophyta: Magniolopsida: Fabales)

Dataset published by Lavin and Sousa (1995)

| | |
|---|---|
| Characters in published dataset: | 37 |
| Taxa in published dataset: | 25 |
| Character relevance cutoff: | 0.95 |
| Characters used to calculate BD: | 37 |
| Taxa used in BDC and MDS analysis: | 25 |
| Stress for 3D MDS: | 0.141 |
| $k_{min}$: | 6 |
| Median bootstrap value: | 71.5 |
| $F_{90}$: | 0.167 |

The legume tribe Robinieae includes the babybonnets, ironwoods, locusts, and the rattlebox. The twelve genera of Robinieae are primarily Neotropical and specifically North American in distribution. Lavin and Sousa (1995) examined the phylogeny of this tribe using morphological characters. Their taxon sample included all twelve Robinieae genera and thirteen legume outgroup taxa. Their character set consisted of 29 reproductive and eight vegetative characters. They found 1224 equally parsimonious trees, in which the monophyly of Robinieae was poorly supported.

All taxa and characters were used to calculate baraminic distances. The BDC results reveal two overlapping groups (Figure 96). The first group contains the genera of Robinieae and the genera *Wisteria* and *Caragana*. The second group overlaps the first with significant, positive BDC in six taxon pairs, involving *Schefflerodendron*, *Millettia*, and *Tephrosia*. Otherwise, the groups are separated by significant, negative BDC. Bootstrap values are exceptionally poor, even within the Robinieae, suggesting that these characters cannot recover even recognized taxonomic groups.

There is one diffuse cluster of taxa in the MDS results (Figure 97), with the two BDC groups on either side of the main cluster. *Wisteria*, *Schefflerodendron*, and *Millettia* occupy the center of the cluster between the Robinieae and the other taxa.

Though the significant, negative BDC between the two BDC groups would imply discontinuity, the clustering pattern in MDS does not support this conclusion. Critical to the baraminological interpretation is the position of *Schefflerodendron* and *Millettia*. They are positively correlated with three of the Robinieae/*Wisteria*/*Caragana* group and four

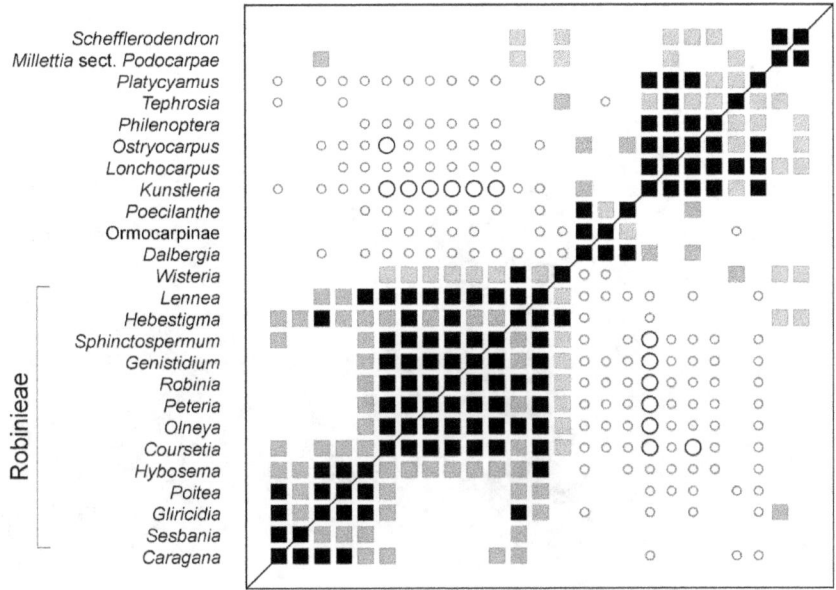

Figure 96. BDC bootstrap results for Robinieae, as calculated by BDISTMDS (relevance cutoff 0.95). Closed square indicate significant, positive BDC; open circles indicate significant, negative BDC. Black symbols indicate bootstrap values >90% in a sample of 100 pseudoreplicates. Grey symbols represent bootstrap values ≤90%.

of the other outgroup taxa. They are negatively correlated with no other taxa, and in the MDS analysis, they are the central taxa between the Robinieae and other outgroup taxa. *Schefflerodendron* and *Millettia* bridge the two BDC groups and thereby imply that the significant, negative BDC is not evidence of discontinuity. Either the entire set of taxa in this study are members of a single holobaramin, or the characters are insufficient to recognize discontinuity. Future studies should include a broader sampling of legume taxa and additional characters.

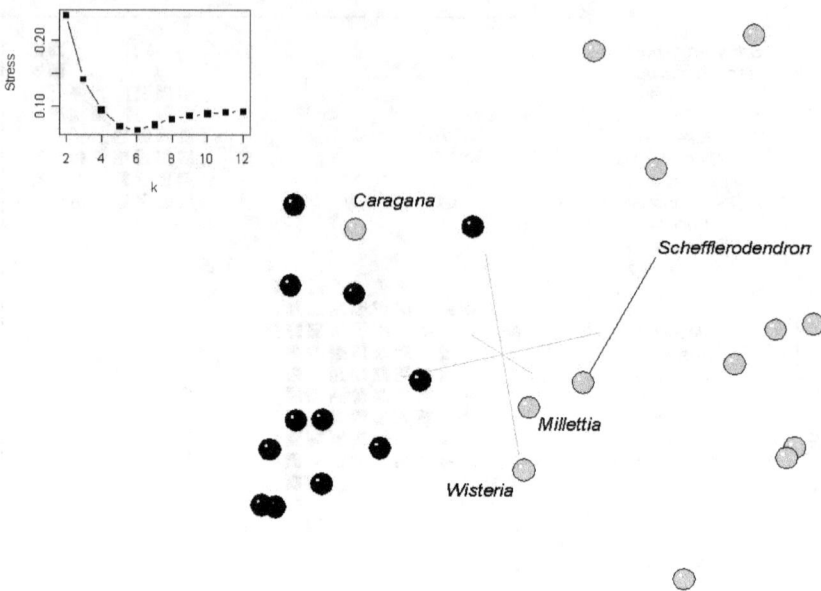

Figure 97. Three-dimensional MDS applied to Robinieae baraminic distances and the stress of k-dimensional MDS on the same baraminic distance matrix plotted as a function of the number of dimensions (k). Robinieae is shown in black and outgroup taxa in gray.

### 3.8. Olacaceae (Magniolophyta: Magniolopsida: Santalales)

Dataset published by Malécot et al. (2004)

| | |
|---|---|
| Characters in published dataset: | 80 |
| Taxa in published dataset: | 59 |
| Character relevance cutoff: | 0.95 |
| Characters used to calculate BD: | 53 |
| Taxa used in BDC and MDS analysis: | 59 |
| Stress for 3D MDS: | 0.225 |
| $k_{min}$: | 8 |
| Median bootstrap value: | 79 |
| $F_{90}$: | 0.331 |

Cronquist (1981, pp.681-684) defined the olax family Olacaceae broadly to include taxa such as the Tanzanian endemic *Octoknema*, which was later placed in its own family. The family was estimated to have 25-30 genera and 250 species. Recent phylogenetic research by Malécot et al. (2004) using morphological data and by Malécot and Nickrent (2008) using nuclear and chloroplast DNA sequences have shown that the "family" is a paraphyletic assemblage of clades at the base of the Santalales. The Santalales are notable for their parasitic and hemiparasitic species, including European mistletoe (*Viscum album*). Their morphological dataset contained a broad sample of 57 taxa in the Santalales and two outgroups from Saxifragales and Caryophyllales. The sample of Santalales included six species of Viscaceae (7-8 genera, 350 species), eleven species of Santalaceae (35 genera, 400 species), six species of Opiliaceae (9 genera, < 50 species), four species of Loranthaceae (60-70 genera, 700 species), one species of Misodendronaceae (1 genus, 10 species), and 29 species of Olacaceae. The Olacaceae taxa represented by 28 different genera. Their characters were based on 39 vegetative anatomical and morphological features (wood, stem, leaf, etc.) and 41 reproductive characteristics.

All taxa and 53 characters from Malécot et al.'s (2004) dataset were used to calculate baraminic distances (11 vegetative and 16 reproductive characters were omitted). There are two groups apparent in the BDC results, and both groups contain taxa assigned to Olacaceae (Figure 98). The smaller group (Group 1) corresponds to the outgroups and Clade 1 of Malécot et al. (2004). Clade 1 contains *Coula, Ochanostachys, Minquartia, Heisteria, Maburea, Scorodocarpus, Strombosia, Strombosiopsis, Tetrastylidium, Engomegorna, Diogoa, Octoknema,* and *Erythropalum*.

The remaining taxa from Olacaceae (14 spp.), Viscaceae, Santalaceae, Opiliaceae, Loranthaceae, and Misodendronaceae comprise the second group. Each group is characterized by extensive positive BDC with high (>90%) bootstrap values. The groups are separated by extensive negative BDC, also with high (>90%) bootstrap values. The outgroup *Daphniphyllum* (Saxifragales: Daphniphyllaceae) partially connects the two groups. Though negatively correlated with 24 members of the larger Group 2, it is also positively correlated with four members of Olacaceae (*Curupira*, *Couradoa*, *Ximenia*, and *Malania*), with bootstrap values of 60-80%.

The presence of two groups of taxa is confirmed in MDS (Figure 99), though the results are not particularly good (3D stress 0.225). The outgroup *Rhabdodendron* occurs between the two groups, and *Daphniphyllum* is an outlier from both groups. Among the Group 2 taxa, *Daphniphyllum* is closest to a small cluster composed of *Curupira*, *Couradoa*, *Ximenia*, and *Malania*, as expected from the BDC results.

Given the present results, there seems to be evidence of only a single discontinuity, between Groups 1 and 2. Within the groups, though, continuity is dubious. Within Group 1, the outgroup taxa *Rhabdodendron* and *Daphniphyllum* are correlated with members of Olacaceae, which seems dubious since they are from three different orders. Continuity within Group 2 could be less dubious, since those taxa are all members of the order Santalales, but the diversity of most of the families (other than Olacaceae) included in Group 2 was only sparsely sampled. Considering past experiences with turtle baraminology (Wood 2005a), it is possible that this dataset contains too many holobaramins, which obscures conclusive recognition of discontinuity. In the case of the turtles, one dataset consisting of a broad representation of turtles revealed only one "discontinuity" in the BDC results, but the MDS showed several more clusters that were only apparent in BDC on datasets composed of a portion of full taxonomic sample. It is possible that with additional taxa from other Santalales families that this dataset would reveal additional discontinuities not evident in the present study. Adding characters to the character sample could be accomplished by additional research to fill in the missing character states for the 27 characters that were omitted from the present study. These also should help to clarify baraminological assignments. Finally, it is also possible that the entire order Santalales is a holobaramin and that the apparent discontinuity detected in this study is an artifact of character sampling. This would result in a holobaramin of about 2000 species, of similar size as other plant holobaramins such as Poaceae.

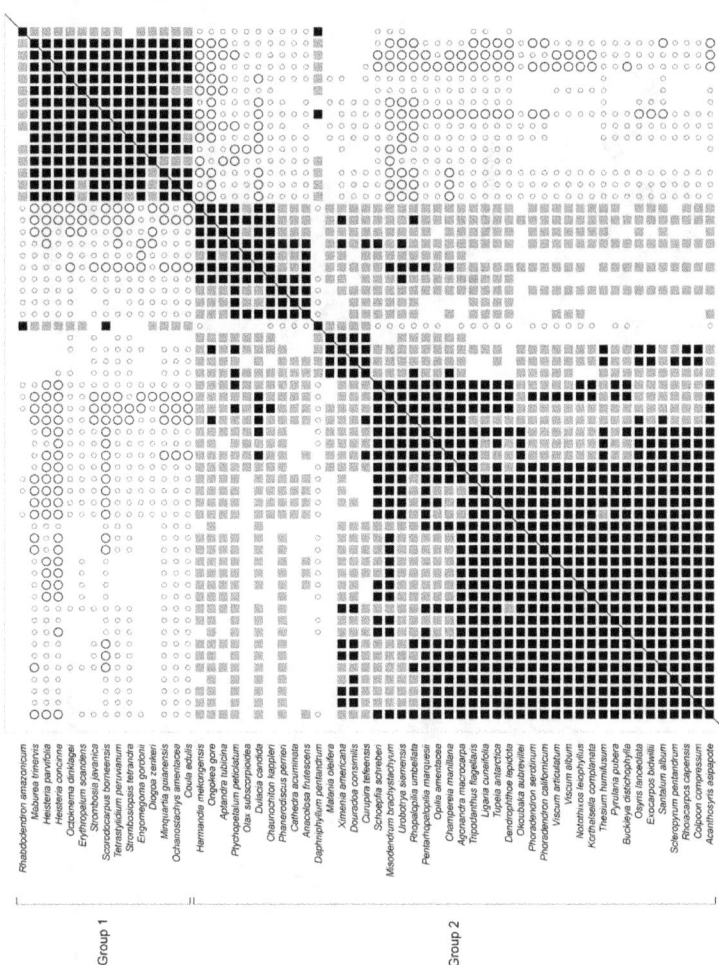

Figure 98. BDC bootstrap results for Olacaceae, as calculated by BDISTMDS (relevance cutoff 0.95). Closed square indicate significant, positive BDC; open circles indicate significant, negative BDC. Black symbols indicate bootstrap values >90% in a sample of 100 pseudoreplicates. Grey symbols represent bootstrap values ≤90%.

# 174 Animal and Plant Baramins

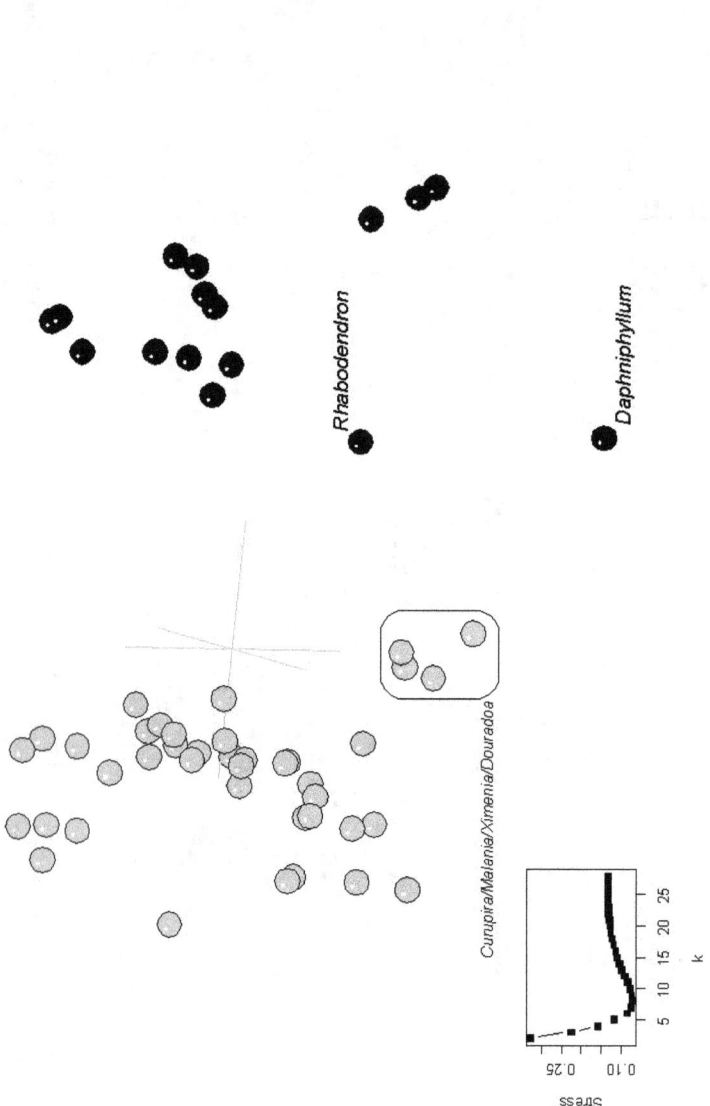

Figure 99. Three-dimensional MDS applied to Olacaceae baraminic distances and the stress of $k$-dimensional MDS on the same baraminic distance matrix plotted as a function of the number of dimensions ($k$). The groups observed in the BDC results (Figure 100) are shown in black and gray.

## 3.9. Celastraceae (Magniolophyta: Magniolopsida: Celastrales)

Dataset published by Simmons et al. (2001)

| | |
|---|---|
| Characters in published dataset: | 61 |
| Taxa in published dataset: | 59 |
| Character relevance cutoff: | 0.95 |
| Characters used to calculate BD: | 42 |
| Taxa used in BDC and MDS analysis: | 59 |
| Stress for 3D MDS: | 0.219 |
| $k_{min}$: | 7 |
| Median bootstrap value: | 66 |
| $F_{90}$: | 0.233 |

Celastraceae is a family of shrubs, trees, and vines found primarily in tropical regions. Conservatively, there are approximately 850 species in 50 genera, but other estimates range as high as 1300 species in 94 genera. There is also disagreement on the inclusion in the Celastraceae *sensu lato* of the tropical Hippocrateaceae. According to Cronquist (1981, p. 716), Hippocrateaceae contains 300 species in two genera. To study the phylogeny of Celastraceae *s. l.*, Simmons et al. (2001) conducted a morphological and molecular analysis. Their morphological dataset consisted of 61 vegetative, reproductive, and cytogenetic characters and 59 taxa. They sampled 49 members of Celastraceae *s.l.* including eight species from Hippocrateaceae. The eight outgroup taxa came from eight different families closely allied with the Celastraceae.

All taxa and 42 of the 61 characters were used to calculate baraminic distances (omitted characters: 27-28, 33, 42, 46-54, 56-61). Correlation analysis reveals a single group of taxa connected by significant, positive BDC but with sporadic negative BDC also (Figure 100). The eight outgroup taxa have significant, positive BDC with 36 different members of Celastraceae. There are 98 instances of significant, negative BDC between outgroup and ingroup taxa, but there are also 169 instances of significant, negative BDC between members of the Celastraceae ingroup. Bootstrap values are overall quite poor, but are highest for significant, positive BDC within Celastraceae. Only thirteen instances of significant, positive BDC between Celastraceae species and outgroups have bootstrap values >90%, and only two instances of significant, negative BDC between two Celastraceae species have bootstrap values >90%. This would suggest that the connections between Celastraceae and the outgroup taxa and the negative BDC within the Celastraceae are both highly dependent on this particular set of 42 characters.

Though the MDS results are poor (3D stress 0.219), the distribution of taxa is somewhat unusual (Figure 101). The Celastraceae do form a single heterogeneous cluster, which could account for the significant, negative BDC, but the outgroup taxa do not form a separate cluster as they do in many other MDS analyses. Instead, the outgroups form an arc that begins with *Crossosoma* (adjacent to *Maytenus* of the Celastraceae s. l.) and ends with *Goupia* (adjacent to *Brexia* and *Mortonia* of the Celastraceae s. l.). The central taxa of the outgroup arc, *Eucryphia*, *Afrostyrax*, *Averrhoa*, are separated from the Celastraceae cluster, but the termini are closely adjacent to some Celastraceae taxa. The closest outgroups to Celastraceae, *Goupia* and *Crossosoma*, have extensive positive BDC with members of the ingroup (19 and 17 instances respectively).

The interpretation of these results is challenging. Outgroups generally form separate clusters or clusters of adjacent taxa. In the case of separate clusters of outgroups, discontinuity may be inferred, but when outgroups are closely intermingled with the ingroups, discontinuity is not apparent. In evolutionary terms, a cluster of outgroups also makes some sense, since usually one taxon will be closest to the ingroup and thereby represent its "sister taxon." The interpretation of an arc with two different taxa closest to two different ingroups is not at all obvious. The phylogenetic results of Simmons et al. (2000) indicate that the outgroup taxa form a distinct clade that is the sister taxon to the entire Celastraceae s. l. The most basal of the Celastraceae are *Mortonia* and *Perrottetia*, only one of which is closely adjacent to an outgroup in the MDS results (*Mortonia* and the outgroup *Goupia*). Though they are nearly indistinguishable in the MDS distribution, *Mortonia* and *Brexia* actually appear on different locations in the Celastraceae phylogeny, with *Brexia* nested well within the Celastraceae clade. *Siphonodon*, despite clustering very closely with the outgroup taxa appears as one of the most derived Celastraceae in the phylogenetic study.

Given these considerations, the baraminology of this group of taxa is difficult to determine. It is possible that all nine families here represented form a single holobaramin. It is also possible that the Celastraceae are actually two different holobaramins, each one adjacent to a different set of outgroups. Celastraceae could also be a holobaramin all by itself (including possibly *Goupia*), but the present character or taxon sampling prevents the recognition of discontinuity. It might be desirable to sample additional members of the outgroup families to evaluate the diversity within the outgroups. Future studies could also work on filling in missing data and thereby expanding the character set from which baraminic distances could be calculated. At present, these results are inconclusive.

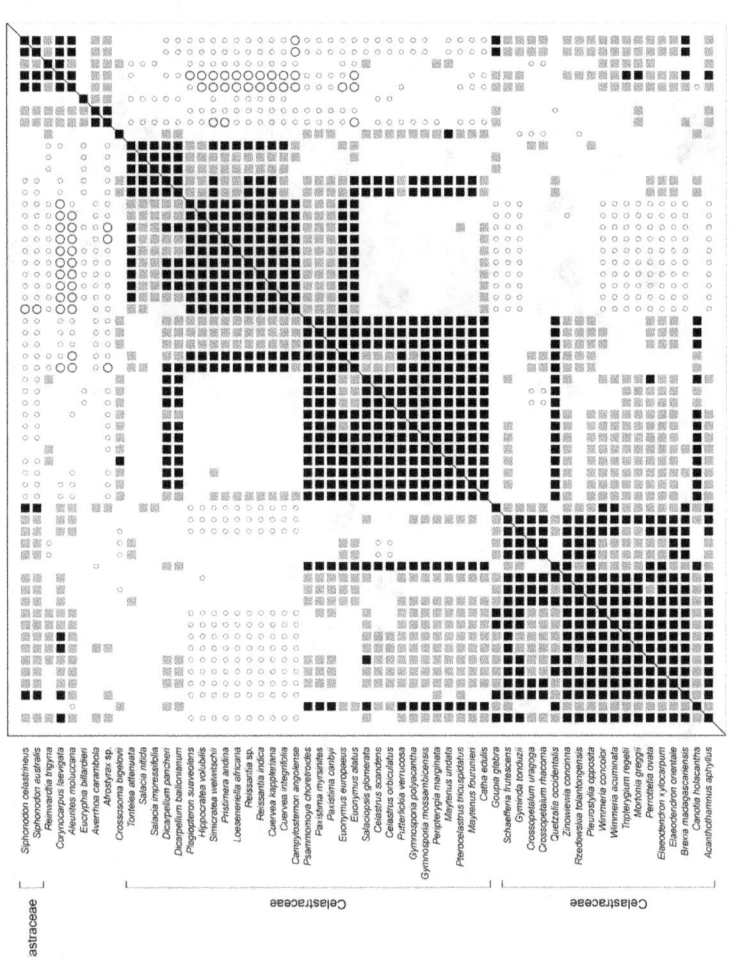

Figure 100. BDC bootstrap results for Celastraceae, as calculated by BDISTMDS (relevance cutoff 0.95). Closed square indicate significant, positive BDC; open circles indicate significant, negative BDC. Black symbols indicate bootstrap values > 90% in a sample of 100 pseudoreplicates. Grey symbols represent bootstrap values ≤90%.

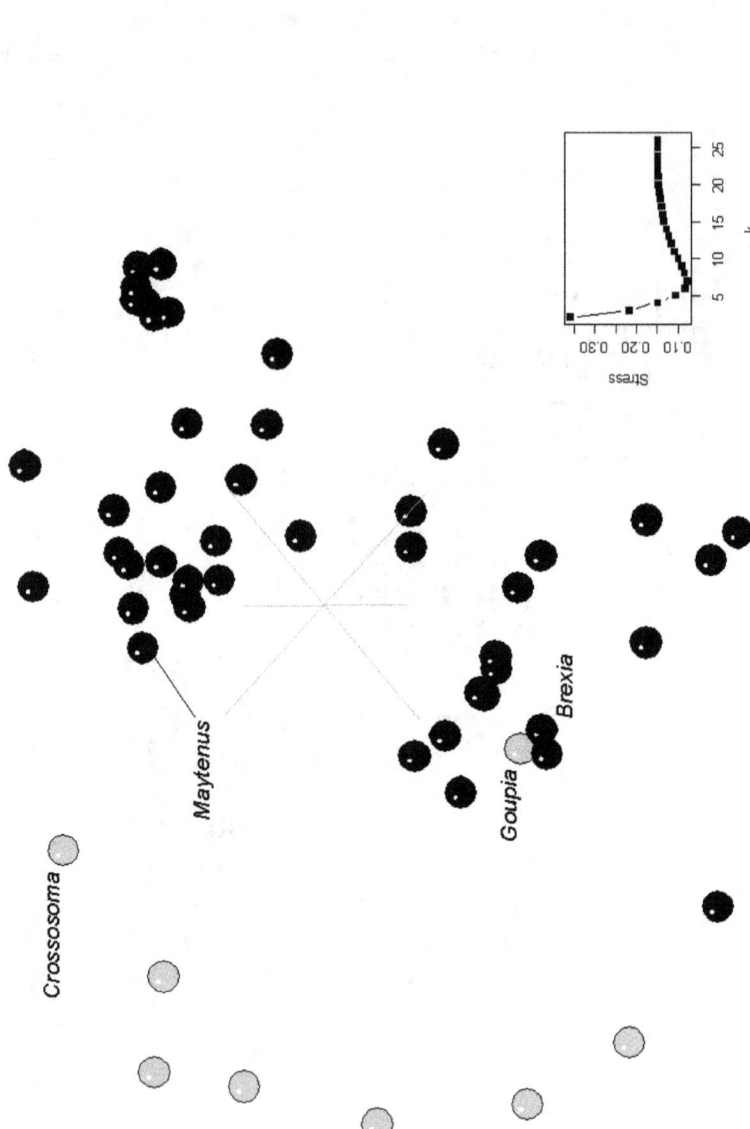

Figure 101. Three-dimensional MDS applied to Celastraceae baraminic distances and the stress of $k$-dimensional MDS on the same baraminic distance matrix plotted as a function of the number of dimensions ($k$). Celastraceae taxa are shown in black and outgroup taxa in gray.

## 3.10. Rubiaceae (Magniolophyta: Magniolopsida: Rubiales)

Dataset published by Bremer (1996)

| | |
|---|---|
| Characters in published dataset: | 35 |
| Taxa in published dataset: | 33 |
| Character relevance cutoff: | 0.95 |
| Characters used to calculate BD: | 35 |
| Taxa used in BDC and MDS analysis: | 33 |
| Stress for 3D MDS: | 0.26 |
| $k_{min}$: | 8 |
| Median bootstrap value: | 73 |
| $F_{90}$: | 0.11 |

The coffee family Rubiaceae is one of the larger flowering plant families with 7,000-10,000 species in 500-600 genera. Found worldwide (including species of *Galium* native to the coasts of Greenland), Rubiaceae species are concentrated in the tropical and subtropical regions. The family is also economically important as the source of coffee and quinine. Bremer's (1996) phylogenetic analysis sampled 33 genera of Rubiaceae, representing 19 of the 44 tribes. No outgroups were included. Her morphological dataset consisted of 35 vegetative, reproductive, chemistry, and cytogenetic characters. Based on six most parsimonious trees, she recognized three main clades of Rubiaceae, corresponding to subfamilies Cinchonoideae, Ixoroideae, and Rubioideae.

All taxa and characters were used to calculate baraminic distances. Only one main group consisting of all taxa was detected in the BDC results (Figure 102). Taxon pairs have sporadic positive BDC with poor bootstrap values. One block of taxa corresponding to the subfamily Rubioideae has signficiant, positive BDC for all 36 taxon pairs within group, and 33 of those taxon pairs have bootstrap values >90%. Significant, negative BDC also occurs sporadically, but only two instances have bootstrap values >90%. MDS results are poor (3D stress 0.26), and as expected, the taxa form a single cluster (Figure 103).

These results probably indicate that the Rubiaceae are a monobaramin. No evidence of discontinuity was observed, thus ruling out any holobaramin identification. Given the small (though diverse) sample of Rubiaceae species, this result should be considered tentative pending further investigation. Future studies should include additional Rubiaceae taxa and some outgroups to test both continuity within and discontinuity around the family. Given the present results, though, the inclusion of these Rubiaceae taxa into a single monobaramin seems likely.

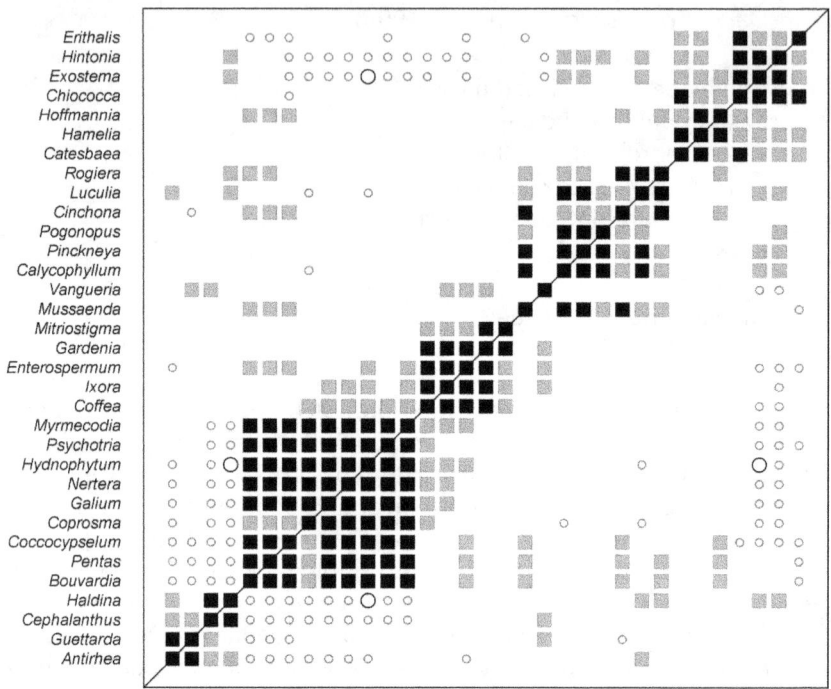

Figure 102. BDC bootstrap results for Rubiaceae, as calculated by BDISTMDS (relevance cutoff 0.95). Closed square indicate significant, positive BDC; open circles indicate significant, negative BDC. Black symbols indicate bootstrap values >90% in a sample of 100 pseudoreplicates. Grey symbols represent bootstrap values ≤90%.

Plants 181

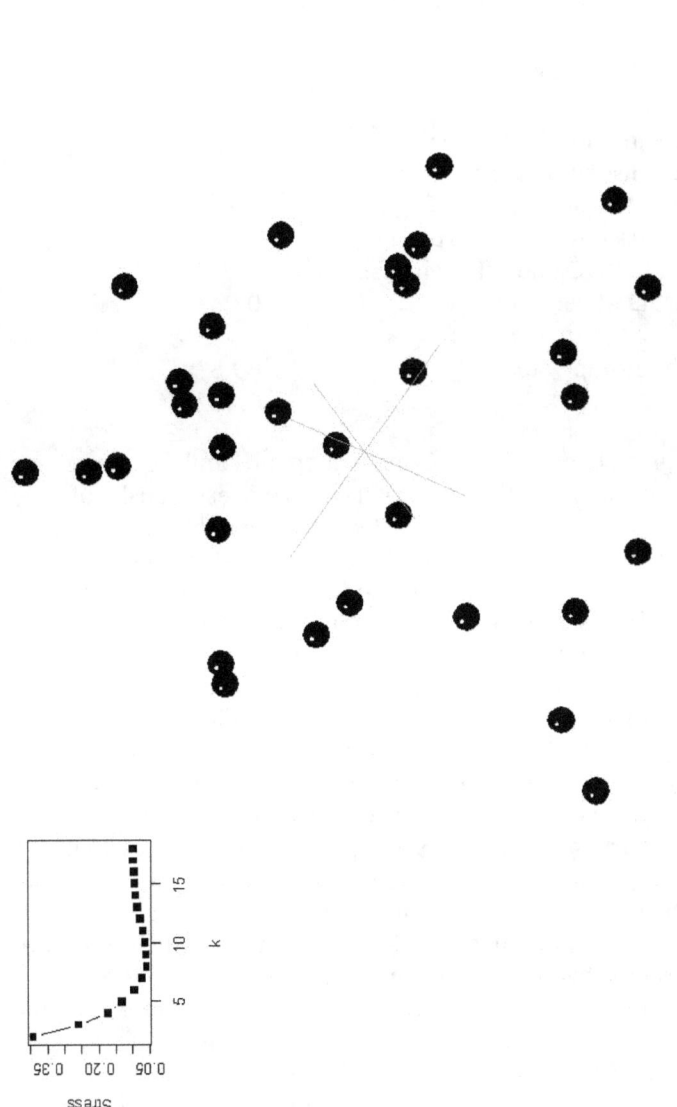

Figure 103. Three-dimensional MDS applied to Rubiaceae baraminic distances and the stress of *k*-dimensional MDS on the same baraminic distance matrix plotted as a function of the number of dimensions (*k*).

### 3.11. Zosteraceae (Magniolophyta: Liliopsida: Najadales)

Dataset published by Les et al. (2002)

| | |
|---|---|
| Characters in published dataset: | 31 |
| Taxa in published dataset: | 13 |
| Character relevance cutoff: | 0.95 |
| Characters used to calculate BD: | 27 |
| Taxa used in BDC and MDS analysis: | 13 |
| Stress for 3D MDS: | 0.043 |
| $k_{min}$: | 4 |
| Median bootstrap value: | 89.5 |
| $F_{90}$: | 0.462 |

The eel grass family Zosteraceae is a small family of monocots that live entirely submerged in sea water. Zosteraceae consist of only eighteen species in three genera (*Zostera*, 12 spp.; *Heterozostera*, 1 sp.; *Phyllospadix*, 5 spp.). Les et al. (2002) evaluated the phylogeny of the family using molecular sequences and vegetative and reproductive morphology. They sampled 31 characters and 13 taxa (11 *Zostera* species, and one each of *Phyllospadix* and *Heterozostera*). They found 68 most parsimonious trees, and the majority rule consensus tree divided the family into three clades: *Zostera* (*Zostera*) including *Heterozostera*, *Zostera* (*Zosterella*), and *Phyllospadix* (which was used as an outgroup for the other taxa). The strict consensus tree did not include *Heterozostera* in the *Zostera* (*Zostera*) clade, but instead had an unresolved trichotomy between *Heterozostera*, *Zostera* (*Zostera*), and *Zostera* (*Zosterella*).

All taxa and 27 characters were used to calculate baraminic distances (omitted characters: 1, 8, 30-31). BDC results revealed three groups with good bootstrap support (Figure 104). The first group includes seven *Zostera* species of subgenus *Zosterella* and *Heterozostera*. The second group contains the four *Zostera* species of subgenus *Zosterella*, and the third group contains only *Phyllospadix*. Within the two multi-species groups, all taxon pairs have significant, positive BDC and only two of those have bootstrap values < 90%. Only four instances of significant, negative BDC occur, all with bootstrap values > 90% and all involving *Phyllospadix* and a member of the *Zostera* (*Zostera*). No significant BDC (positive or negative) occurs between *Zostera* (*Zostera*) and *Zostera* (*Zosterella*) or between *Phyllospadix* and *Zostera* (*Zosterella*). *Heterozostera* is positively correlated with all members of *Zostera* (*Zosterella*), but has no significant BDC with any other taxa.

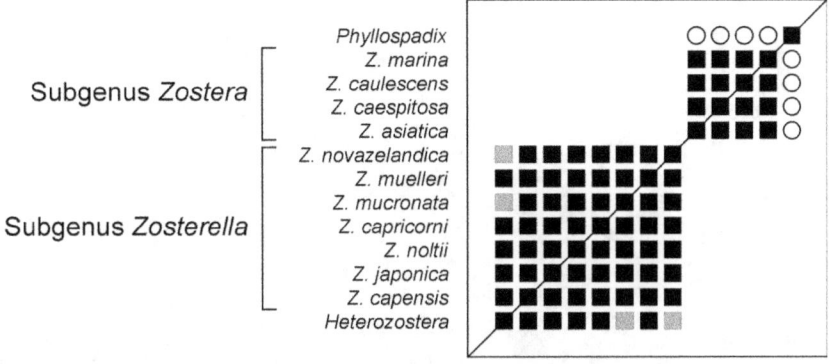

Figure 104. BDC bootstrap results for Zosteraceae, as calculated by BDISTMDS (relevance cutoff 0.95). Closed square indicate significant, positive BDC; open circles indicate significant, negative BDC. Black symbols indicate bootstrap values >90% in a sample of 100 pseudoreplicates. Grey symbols represent bootstrap values ≤90%.

The MDS results are extremely good (3D stress 0.043) and correspond very well to the BDC results. Two clusters of taxa and one distant outlier are apparent, corresponding precisely to the three groups identified in the BDC results (Figure 105). They form an irregular triangle, with *Zostera* (*Zostera*) and *Phyllospadix* being the most distant. In between is *Zostera* (*Zosterella*) with *Heterozostera* slightly offset.

These results are unlikely to tell us anything about the baraminology of Zosteraceae. The extremely good clustering seen in the MDS is more likely the result of biased character selection that emphasizes differences between the genera and subgenera. This also seemed to be the case in the Sulidae analysis of Wood (2005a), where four genera of taxa were shown to be distributed in a nearly regular tetrahedron, indicating that the genera were roughly equidistant. Here, the genera and subgenera are not equidistant but are nevertheless very distinct. As in the case of the sulids, the BDC results show positive correlation within each genus/subgenus and little significant correlation between the genera/subgenera. Future research on the baraminology of Zosteraceae should focus on a different sample of characters and should include outgroup taxa.

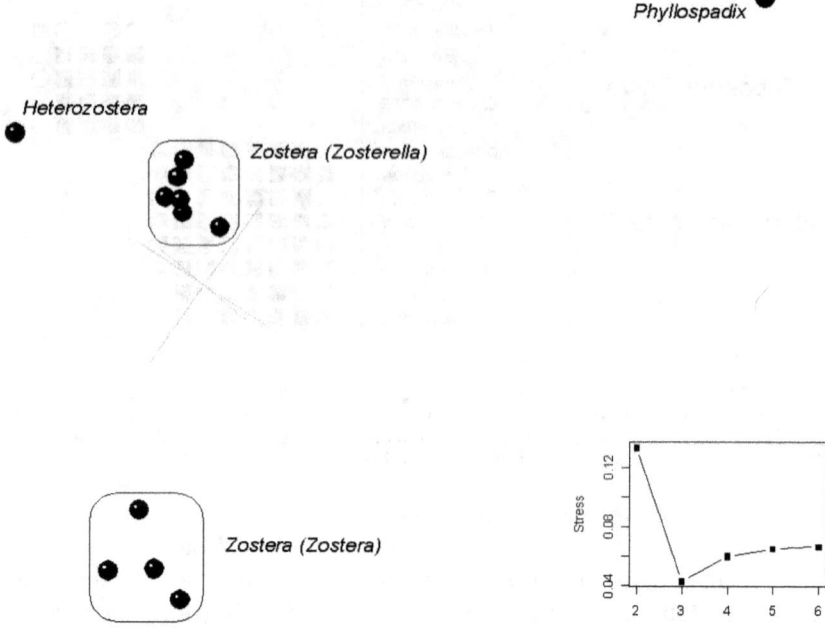

Figure 105. Three-dimensional MDS applied to Zosteraceae baraminic distances and the stress of *k*-dimensional MDS on the same baraminic distance matrix plotted as a function of the number of dimensions (*k*).

## 3.12. Lemnaceae (Magnoliophyta: Liliopsida: Arales)

Dataset published by Stockey et al. (1997)

| | |
|---|---|
| Characters in published dataset: | 34 |
| Taxa in published dataset: | 12 |
| Character relevance cutoff: | 0.9 |
| Characters used to calculate BD: | 32 |
| Taxa used in BDC and MDS analysis: | 12 |
| Stress for 3D MDS: | 0.056 |
| $k_{min}$: | 4 |
| Median bootstrap value: | 92 |
| $F_{90}$: | 0.53 |

The duckweeds are the smallest flowering plants. Tiny *Wolffia* grows to just under half a millimeter. The typical duckweed consists of a single "leaf" or thallus that floats in great numbers on any still freshwater. Other tissues (stems, roots, etc.) have been greatly reduced or are absent. Reproduction is usually asexual; flowers are rarely produced. There are approximately 40 species of duckweeds in four genera. Cronquist (1981) allied the Lemnaceae with the arum family Araceae, especially with the water cabbage *Pistia stratiotes*. Stockey et al.'s (1997) phylogenetic study indicated that Araceae are actually paraphyletic with the Lemnaceae, with *Pistia* being the most closely-related of the "Araceae" to the Lemnaceae. They also found that the Miocene fossil *Limnobiophyllum* was either the sister taxon to the Lemnaceae or the most basal of the Lemnaceae. Their phylogenetic analysis was based on 34 characters covering vegetative and reproductive attributes as well as natural history characteristics. Their taxon sample consisted of all four duckweed genera, *Limnobiophyllum*, six genera of Araceae (including *Pistia*) and the outgroup *Acorus* (placed either in Araceae or its own family Acoraceae).

All twelve taxa and 32 of 34 characters were used to calculate baraminic distances (omitted characters: 17 and 34). BDC results reveal two basic groups: Araceae (including *Acorus* and *Pistia*) and Lemnaceae (including *Limnobiophyllum*) (Figure 106). Bootstrap values are highest (>90%) for taxon pairs within the Araceae. Three taxa are connected to the groups by one or two significant, positive BDC: *Pistia* (to *Colocasia*), *Limnobiophyllum* (to *Spirodella*), and *Acorus* (to *Ambrosina* and *Cyrtosperma*). Each of these taxon pairs also has poor bootstrap values (45-83%), indicating that the correlations are dependent on the particular character set chosen. The main two groups are separated by significant, negative BDC with high (>90%) bootstrap values, but there is no significant, positive BDC connecting the two groups.

MDS results reveal a diffuse, arc-like distribution of taxa with Lemnaceae at one end and *Acorus* at the other (Figure 107). Between the Lemnaceae and Araceae is *Limnobiophyllum*, but it is unclear from the MDS results alone to which group it belongs. Adjacent to *Limnobiophyllum* is *Pistia*, which is likewise not obviously part of the Araceae. *Acorus* is similarly set off from the Araceae. Despite these widely-separated taxa, the 3D MDS is of exceptional quality, with a stress of only 0.056.

Given the present results, the Lemnaceae (tentatively including *Limnobiophyllum*) can be identified as a holobaramin, and the Araceae can very tentatively also be identified as a holobaramin. Of these two holobaramins, the Lemnaceae is the more sure classification. All genera of Lemnaceae are here sampled, and discontinuity from the Araceae can be established based on the significant, negative BDC with high bootstrap values. The position of *Limnobiophyllum* is more tentative, since it is based on only a single positive BDC with a poor bootstrap value. The assignment of Araceae as a holobaramin is much more doubtful based on the present data. Araceae consists of minimally 100 genera and 2400 species, of which only six genera are sampled here. A much fuller sample of Araceae would be necessary to validate its holobaraminic status. Given the phylogenetic results of Stockey et al. (1997), which showed that Araceae was paraphyletic with Lemnaceae, it is also possible that both families could together comprise a single holobaramin. Future baraminology studies in this group should focus carefully on a wider sample of Araceae.

Plants 187

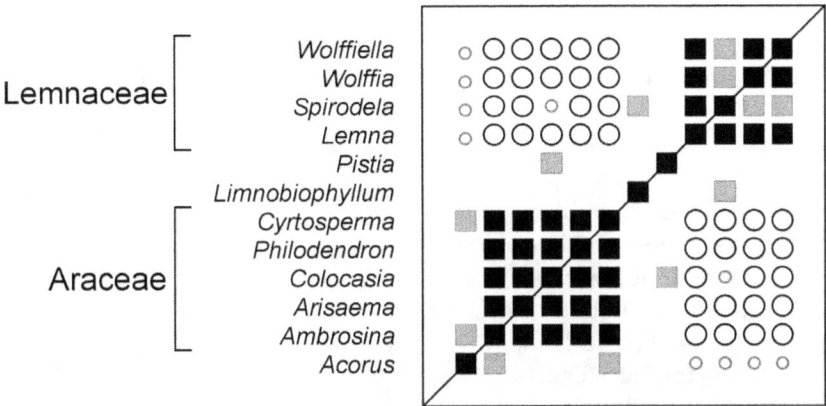

Figure 106. BDC bootstrap results for Lemnaceae, as calculated by BDISTMDS (relevance cutoff 0.9). Closed square indicate significant, positive BDC; open circles indicate significant, negative BDC. Black symbols indicate bootstrap values >90% in a sample of 100 pseudoreplicates. Grey symbols represent bootstrap values ≤90%.

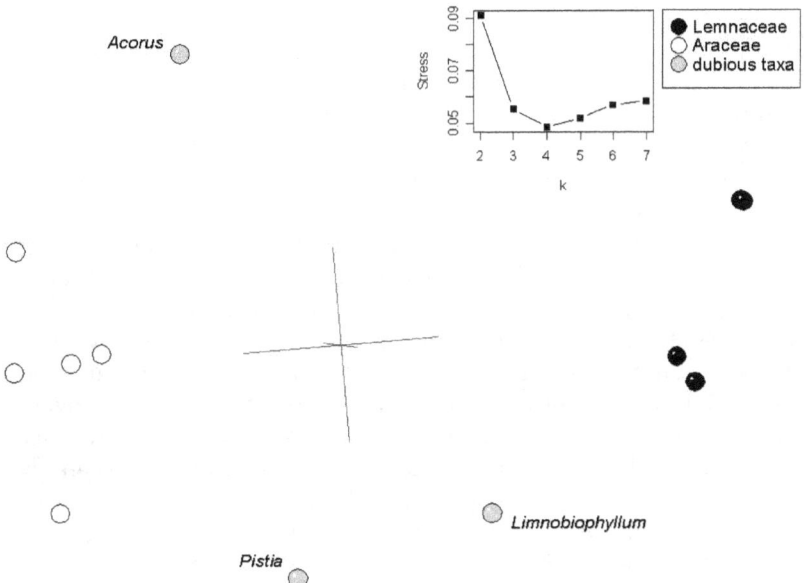

Figure 107. Three-dimensional MDS applied to Lemnaceae baraminic distances and the stress of $k$-dimensional MDS on the same baraminic distance matrix plotted as a function of the number of dimensions ($k$). Lemnaceae is shown in black, Araceae in white, and questionable taxa in gray.

### 3.13. Commelinaceae (Magniolophyta: Liliopsida: Commelinales)

Dataset published by Evans et al. (2000)

| | |
|---|---|
| Characters in published dataset: | 47 |
| Taxa in published dataset: | 42 |
| Character relevance cutoff: | 0.9 |
| Characters used to calculate BD: | 43 |
| Taxa used in BDC and MDS analysis: | 42 |
| Stress for 3D MDS: | 0.197 |
| $k_{min}$: | 7 |
| Median bootstrap value: | 67 |
| $F_{90}$: | 0.226 |

The spiderwort family Commelinaceae consists of about 600 species in 40 genera, native to tropical and subtropical regions. The family has little agricultural importance, but several species are cultivated for ornamentals. In their morphological study of Commelinaceae phylogeny, Evans et al. (2000) included 39 of the recognized genera and one undescribed genus of Commelinaceae, together with two outgroup taxa, *Heteranthera* (Pontederiaceae) and *Haemodorum* (Haemodoraceae). They analyzed 47 largely floral or reproductive characters. They found 154 most parsimonious trees and 80% bootstrap support for the monophyly of the family. Traditional tribal classifications were not found to be monophyletic.

All taxa and 43 of the 47 characters were used for baraminic distance calculations (omitted characters: 29, 34, 39, 42). There is apparently a single group in the BDC results (Figure 108), comprising both Commelinaceae and outgroup taxa. Significant, positive BDC within the Commelinaceae is sporadic, and 134 instances of significant, negative BDC also occur within the family. Bootstrap values are overall quite poor, but bootstrap values >90% for significant, positive BDC are common within the Commelinaceae. The outgroup taxa share five instances of significant, positive BDC with four taxa of Commelinaceae (*Spatholirion*, *Callisia*, *Cartonema*, and *Triceratella*) and fifteen instances of significant, negative BDC with nine taxa of Commelinaceae. None of the outgroup/Commelinaceae correlations have bootstrap values >90%.

The MDS results are poor (3D stress 0.197), but the clustering of taxa is strikingly different from the BDC results. The outgroup taxa are clearly separated from the main, diffuse cluster of Commelinaceae

(Figure 109). *Triceratella*, the only Commelinaceae that is positively correlated with both outgroups, is one of the nearest of the main cluster to the outgroups but is still a definite member of the Commelinaceae cluster. The Commelinaceae cluster is elongated, which explains the significant, negative BDC within the family.

Given the clustering pattern in the MDS results alone, it would seem that Commelinaceae is a holobaramin, separate from the two outgroup taxa. However, given the poor quality of the 3D MDS results and the lack of support from BDC, this conclusion can only be considered very tentative. With most Commelinaceae genera represented, the taxon sampling seems to be adequate, but the entirely reproductive characters may be a biased sampling. Future studies should try to add more vegetative characters to the dataset in order to clarify the holobaraminic status of Commelinaceae.

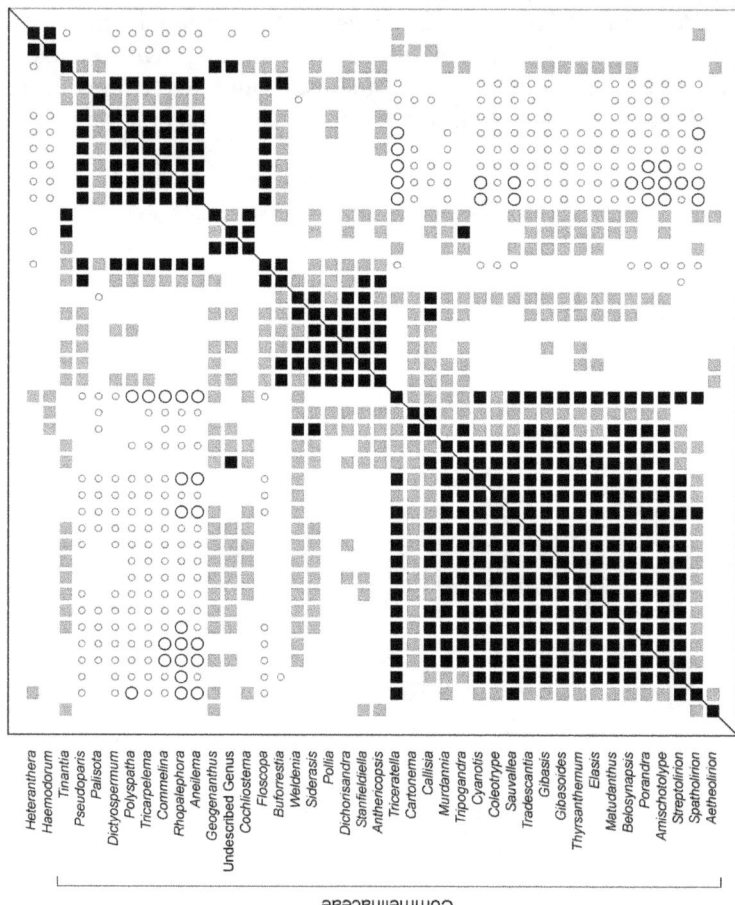

Figure 108. BDC bootstrap results for Commelinaceae, as calculated by BDISTMDS (relevance cutoff 0.9). Closed square indicate significant, positive BDC; open circles indicate significant, negative BDC. Black symbols indicate bootstrap values > 90% in a sample of 100 pseudoreplicates. Grey symbols represent bootstrap values ≤90%.

Plants 191

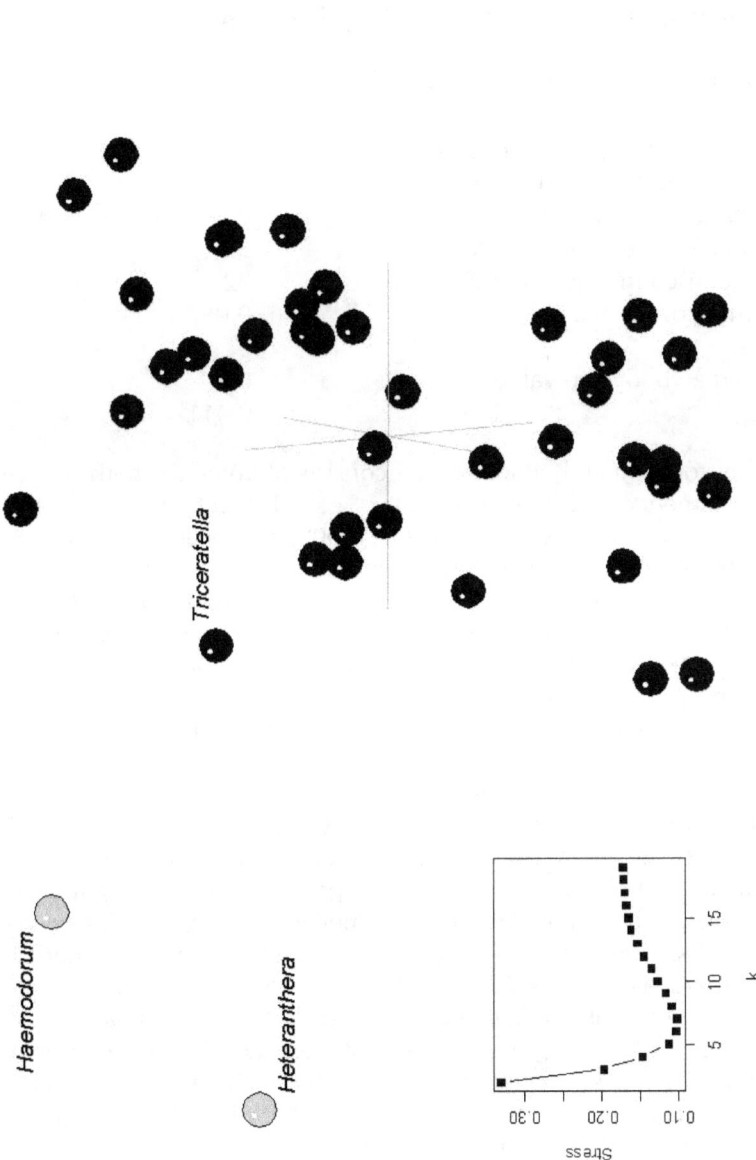

Figure 109. Three-dimensional MDS applied to Commelinaceae baraminic distances and the stress of $k$-dimensional MDS on the same baraminic distance matrix plotted as a function of the number of dimensions ($k$). Commelinaceae taxa are shown in black, outgroup taxa in gray.

### 3.14. Rapateaceae (Magniolophyta: Liliopsida: Commelinales)

Dataset published by Givnish et al. (2000)

| | |
|---|---|
| Characters in published dataset: | 44 |
| Taxa in published dataset: | 22 |
| Character relevance cutoff: | 0.95 |
| Characters used to calculate BD: | 32 |
| Taxa used in BDC and MDS analysis: | 22 |
| Stress for 3D MDS: | 0.097 |
| $k_{min}$: | 6 |
| Median bootstrap value: | 87 |
| $F_{90}$: | 0.411 |

The tropical family Rapateaceae consists of perennial herbs native to South America and west Africa. It is a small family, with only 100 species in sixteen genera. Givnish et al. (2000) examined the phylogeny of Rapateaceae using molecular and morphological characters. Their morphological character set was derived from vegetative and reproductive characteristics. They sampled 17 Rapateaceae species representing fourteen genera. Their outgroups were four species of bromeliads and the bogmoss *Mayaca* of the monogeneric family Mayacaceae. Their morphological and molecular phylogenies supported the monophyly of Rapateaceae.

To calculate baraminic distances, all taxa and 32 of the characters were used (omitted characters: 8-9, 14, 16, 18-22, 26-27, 29). BDC results were nearly identical to the BDC results for an ideal holobaramin. The Rapateaceae and *Mayaca* form a group united by significant, positive BDC and separated from the bromeliad outgroups by significant, negative BDC (Figure 110). No significant, negative BDC was observed within the Rapateaceae + *Mayaca* group. Bootstrap results were quite good, with an overall median of 87%. There are 153 taxon pairs in the Rapateaceae + *Mayaca* group, and 133 of them had significant, positive BDC, with 71 having bootstrap values >90%. There are 72 possible taxon pairs involving a bromeliad and a member of the Rapateaceae + *Mayaca* group, and 70 of these had significant, negative BDC, with 28 having bootstrap values >90%.

The MDS results were also quite good (3D stress 0.097), showing three groups of taxa, corresponding to the three families in the study: Rapateaceae, Bromeliaceae, and Mayacaceae (Figure 111). The Rapateaceae cluster is slightly diffuse, and surprisingly *Mayaca* is not

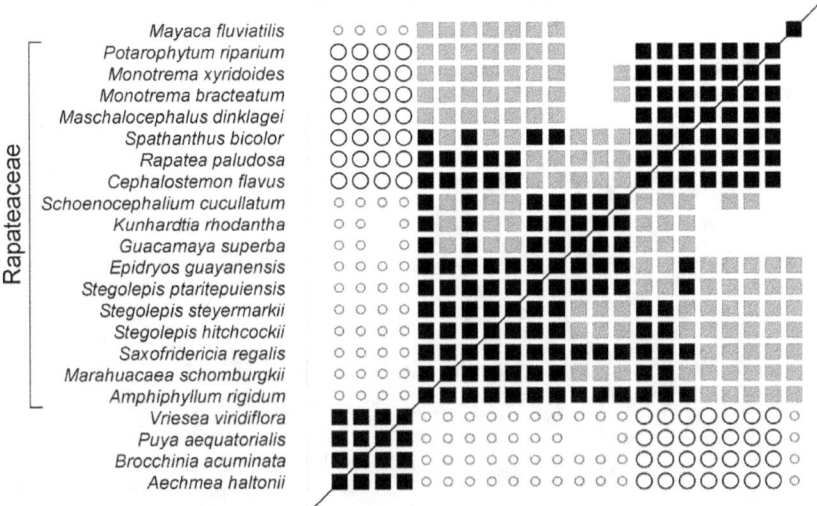

Figure 110. BDC bootstrap results for Rapateaceae, as calculated by BDISTMDS (relevance cutoff 0.95). Closed square indicate significant, positive BDC; open circles indicate significant, negative BDC. Black symbols indicate bootstrap values >90% in a sample of 100 pseudoreplicates. Grey symbols represent bootstrap values ≤90%.

part of the main cluster of Rapateaceae but lies at some distance from it. The Bromeliaceae taxa form an almost perfect linear array in the MDS results.

The BDC results alone support a holobaramin of Rapateaceae + *Mayaca*, with almost universal continuity within the group (as evidenced by significant, positive BDC) and discontinuity with the outgroups (as evidenced by significant, negative BDC). The MDS results suggest that *Mayaca* might be in a separate baraminic group due to its position as an outlier from the Rapateaceae. Nevertheless, given the present data, a holobaramin consisting of both Rapateaceae and Mayacaceae is supported.

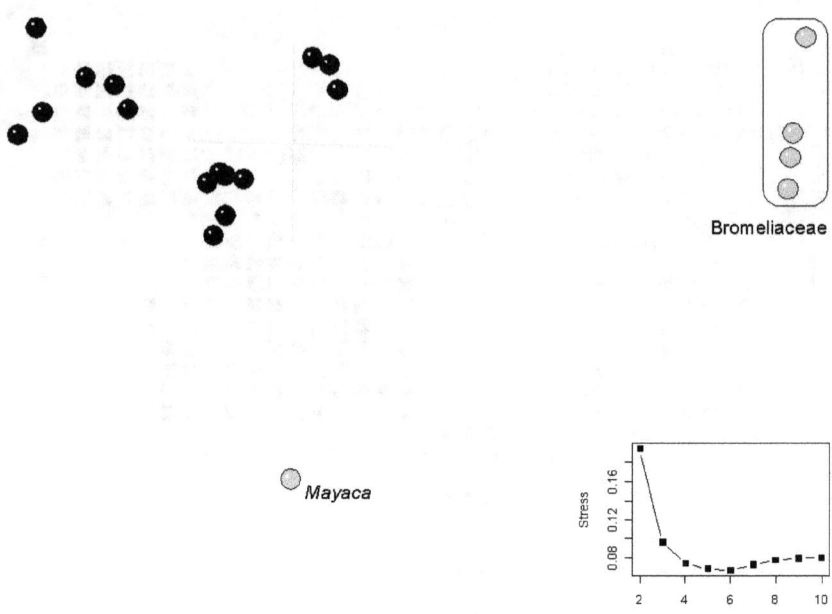

Figure 111. Three-dimensional MDS applied to Rapateaceae baraminic distances and the stress of $k$-dimensional MDS on the same baraminic distance matrix plotted as a function of the number of dimensions ($k$). Rapateaceae is shown in black, outgroup taxa in gray.

### 3.15. Alstroemeriaceae (Magniolophyta: Liliopsida: Liliales)

Dataset published by Aagesen and Sanso (2003)

| | |
|---|---|
| Characters in published dataset: | 61 |
| Taxa in published dataset: | 33 |
| Character relevance cutoff: | 0.95 |
| Characters used to calculate BD: | 48 |
| Taxa used in BDC and MDS analysis: | 33 |
| Stress for 3D MDS: | 0.169 |
| $k_{min}$: | 6 |
| Median bootstrap value: | 100 |
| $F_{90}$: | 0.748 |

The small, lily-like family of Alstroemeriaceae consists of about 200 species in three genera. Members of genus *Alstroemeria*, called Peruvian lilies, are cultivated for ornamental purposes; the species are native to Central and South America. Aagesen and Sanso's (2003) phylogenetic study of the family used molecular sequences and 61 vegetative, reproductive, and cytogenetic characters. They sampled 27 species of Alstroemeriaceae from three genera and six outgroup species from Smilacaceae (three *Luzuriaga* spp.), Uvulariaceae (*Uvularia*), Colchicaceae (*Disporum*), and Dioscoreaceae (*Dioscorea*). Their results supported the monophyly of Alstroemeriaceae and indicated that a clade of *Uvularia* and *Disporum* formed the sister group to the family.

For baraminic distance calculations, all taxa and 48 characters were used (omitted characters: 14, 17-18, 42, 48, 51-52, 56-61). The BDC results show two groups corresponding to the Alstroemeriaceae and the outgroups (Figure 112). Most taxa within the groups have significant, positive BDC, and there are 155 instances of significant, negative BDC between the two groups. The Alstroemeriaceae genus *Leontochir* is correlated with only one other taxon: positively with *Bomarea edulis*. The bootstrap values are extremely high (median 100%), although most intergeneric comparisons in the Alstroemeriaceae have bootstrap values <90%. Nevertheless, 75% of the 528 taxon pairs in this study have bootstrap values >90%.

The MDS results are not as good as the BDC bootstrap results (3D stress 0.169), but the BDC results are generally confirmed (Figure 113). There are two main groups of taxa, a heterogeneous distribution of outgroups and cluster of Alstroemeriaceae. *Leontochir* is an outlier from the main Alstroemeriaceae, but the BDC results suggest that it is tentatively a part of the Alstroemeriaceae and not part of the outgroup. The genus *Bomarea* is also slightly set off from the *Alstroemeria* species, but evidently not enough to eliminate the positive BDC.

These results strongly indicate that Alstroemeriaceae is a holobaramin. The family is united by continuity (evidenced by significant, positive BDC) and separated from other taxa by discontinuity (evidenced by signficiant, negative BDC). The extremely high bootstrap values indicate that this conclusion is relatively impervious to random perturbations in the character set used to calculate baraminic distances. The MDS also supports this general conclusion by showing a cluster of Alstroemeriaceae species separate from the outgroups.

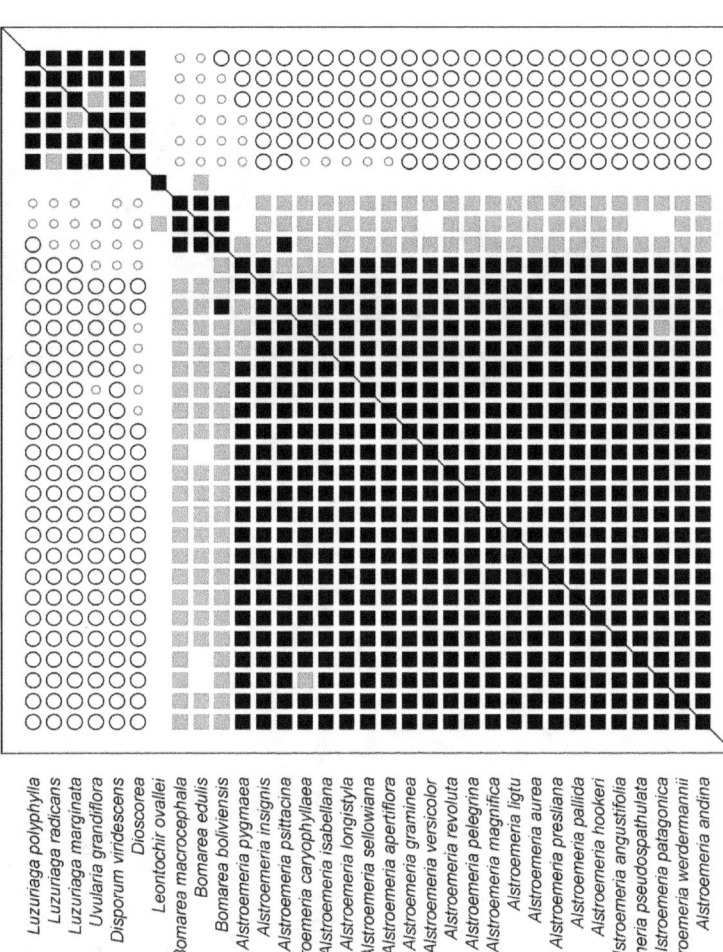

Figure 112. BDC bootstrap results for Alstroemeriaceae, as calculated by BDISTMDS (relevance cutoff 0.95). Closed square indicate significant, positive BDC; open circles indicate significant, negative BDC. Black symbols indicate bootstrap values > 90% in a sample of 100 pseudoreplicates. Grey symbols represent bootstrap values ≤ 90%.

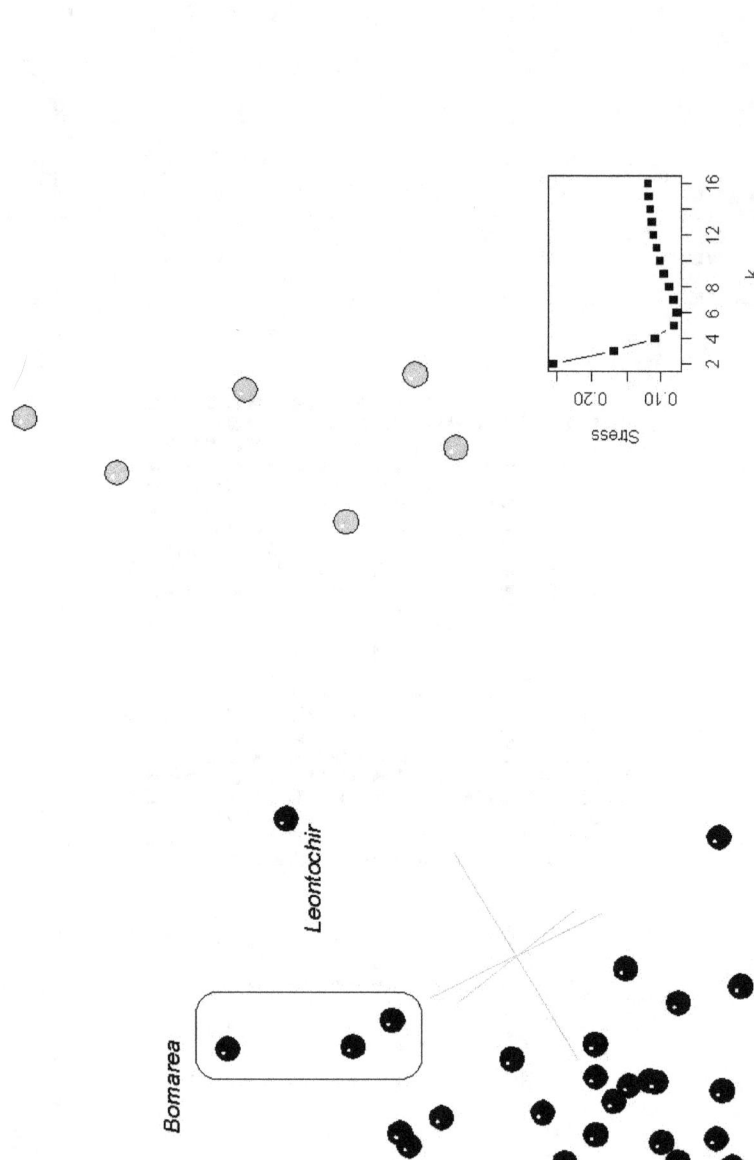

Figure 113. Three-dimensional MDS applied to Alstroemeriaceae baraminic distances and the stress of $k$-dimensional MDS on the same baraminic distance matrix plotted as a function of the number of dimensions ($k$). Alstroemeriacae is shown in black and outgroup taxa in gray.

### 3.16. Pontederiaceae (Magniolophyta: Liliopsida: Liliales)

Dataset published by Graham et al. (1998)

| | |
|---|---|
| Characters in published dataset: | 43 |
| Taxa in published dataset: | 25 |
| Character relevance cutoff: | 0.95 |
| Characters used to calculate BD: | 41 |
| Taxa used in BDC and MDS analysis: | 25 |
| Stress for 3D MDS: | 0.189 |
| $k_{min}$: | 7 |
| Median bootstrap value: | 73 |
| $F_{90}$: | 0.147 |

Pontederiaceae, or the water hyacinth family, contains 30-40 species in eight genera. These species are aquatic and found in tropical and subtropical regions, most frequently in tropical America. Graham et al. (1998) evaluated the phylogeny of the family using molecular sequences and morphological characters. Their character set was updated from Eckenwalder and Barrett (1986) and consisted of 42 natural history, vegetative, and reproductive characters. The taxon sampling included 24 Pontederiaceae species and varieties from five genera and one outgroup *Philydrum* (Philydraceae).

All taxa and 41 characters were used to calculate baraminic distances (omitted: character 32, stamen number). BDC results are an intermixture of positive and negative BDC with poor bootstrap values (median 73%) (Figure 114). Within the Pontederiaceae, only 73 of the 276 possible taxon pairs had significant, positive BDC, and only 30 of those had bootstrap values >90%. There are also 41 instances of significant, negative BDC, and 14 of those had bootstrap values >90%. The outgroup *Philydrum* is poorly distinguished from the Pontederiaceae. There is no significant BDC between *Philydrum* and 22 members of Pontederiaceae. The remaining two Pontederiaceae taxa, *Eichhornia heterosperma* and *E. diversifolia*, were negatively correlated with *Philydrum* with bootstrap values of 48% and 54% respectively.

The MDS results are not very good (3D stress 0.189), but the results generally corroborate the BDC findings. Pontederiaceae is a diffuse cluster of taxa, but *Philydrum* is definitely an outlying taxon from the Pontederiaceae (Figure 115).

From the MDS results, Pontederiaceae could be a holobaramin, with *Philydrum* as the outgroup. The BDC results do not contradict

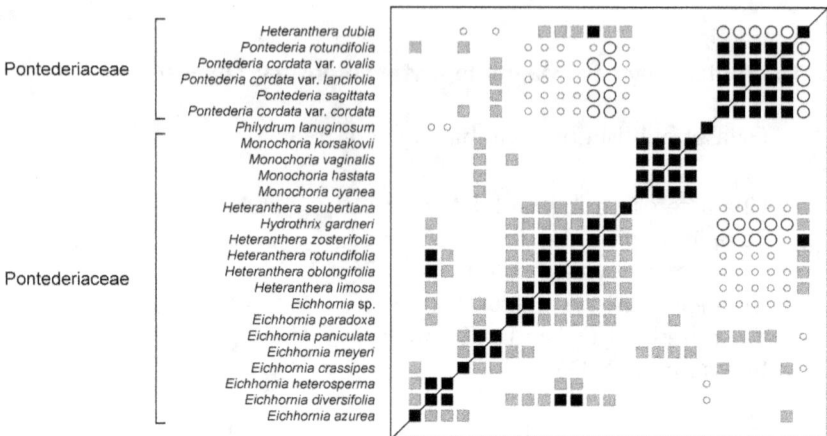

Figure 114. BDC bootstrap results for Pontederiaceae, as calculated by BDISTMDS (relevance cutoff 0.95). Closed square indicate significant, positive BDC; open circles indicate significant, negative BDC. Black symbols indicate bootstrap values >90% in a sample of 100 pseudoreplicates. Grey symbols represent bootstrap values ≤90%.

this conclusion, but they do not support it either. *Philydrum* is neither positively nor negatively correlated with most Pontederiaceae taxa. Pontederiaceae taxa do appear to be monobaraminic though. The significant, positive BDC within the group is scant but does connect most taxa, and the diffuse nature of the Pontederiaceae cluster could account for the lack of positive BDC and the occurrence of sporadic negative BDC. Future studies could expand the character set or add additional outgroup taxa to evaluate the hypothesis of discontinuity between the Pontederiaceae and outgroup taxa.

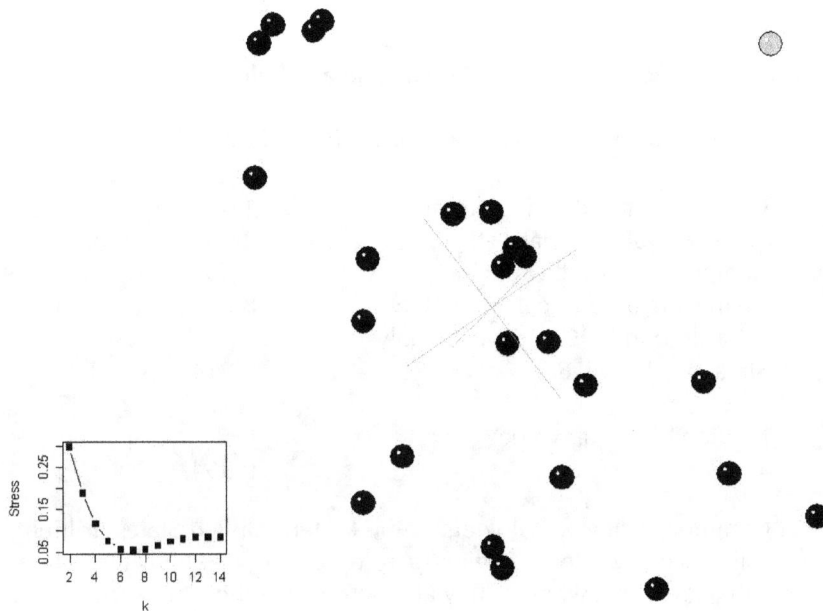

Figure 115. Three-dimensional MDS applied to Pontederiaceae baraminic distances and the stress of $k$-dimensional MDS on the same baraminic distance matrix plotted as a function of the number of dimensions ($k$). Pontederiaceae is shown in black and the outgroup taxon in gray.

### 3.17. Trilliaceae (Magniolophyta: Liliopsida: Liliales)

Dataset published by Farmer and Schilling (2002)

| | |
|---|---|
| Characters in published dataset: | 110 |
| Taxa in published dataset: | 26 |
| Character relevance cutoff: | 0.95 |
| Characters used to calculate BD: | 84 |
| Taxa used in BDC and MDS analysis: | 26 |
| Stress for 3D MDS: | 0.214 |
| $k_{min}$: | 7 |
| Median bootstrap value: | 86 |
| $F_{90}$: | 0.354 |

Common spring wildflowers in the eastern United States, trilliums are easily recognizable by their tripartite appearance, with three large leaves that grow in a whorl. They are members of the family Trilliaceae, which includes approximately 56 species in 4-6 genera. Trilliaceae have been classified variously in the Liliaceae or Melanthiaceae. Farmer and Schilling (2002) examined the relationships of the genera using a dataset of 110 characters and 26 taxa. Their characters came from vegetative, reproductive, cytogenetic, and natural history attributes. Their sampling of taxa for the morphological phylogeny did not included non-Trilliaceae outgroups.

To calculate baraminic distances, all taxa and 84 characters were used (omitted characters: 17, 37, 39, 64-65, 67, 72-73, 77, 79-82, 85-86, 88-89, 93-95, 97-98, 100, 105-106, 109). BDC results show two groups, one containing the genera *Trillium*, *Pseudotrillium*, and *Trillidium* and the other containing the genera *Kinugasa*, *Daiswa*, and *Paris* (Figure 116). Within each group, significant, positive BDC is common, connecting all taxa. Bootstrap values are also quite high (>90%) for these correlations. Between the two groups, 127 of the 144 taxon pairs have significant, negative BDC but only seventeen of those correlations have bootstrap values >90%. No significant, positive BDC occurs between the two groups.

In the MDS results, the two groups from the BDC results appear as separate, orthogonal linear distributions (Figure 117), as also seen in the case of the phalacrocoracids (Wood 2005a). The 3D stress is quite high, however (0.214), indicating that the distribution of taxa is not ideal. There is a separation between the two groups seen in the BDC

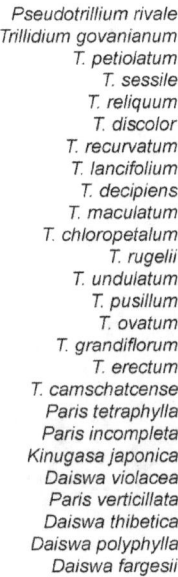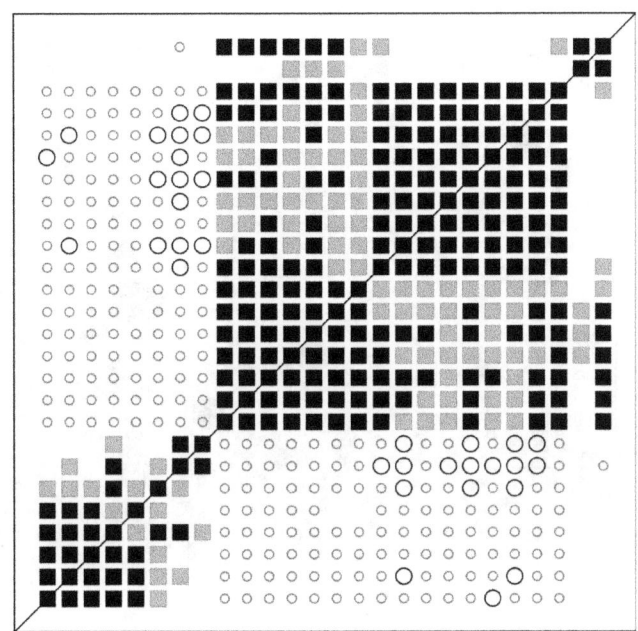

Figure 116. BDC bootstrap results for Trilliaceae, as calculated by BDISTMDS (relevance cutoff 0.95). Closed square indicate significant, positive BDC; open circles indicate significant, negative BDC. Black symbols indicate bootstrap values >90% in a sample of 100 pseudoreplicates. Grey symbols represent bootstrap values ≤90%.

results, but the negative BDC is likely the result of a combination of the separation of the taxa and the peculiar geometry.

It is possible that one or both of the groups observed in the BDC are holobaramins, which seems likely just from the BDC results. However, the orthogonality of the distribution of taxa in the MDS analysis suggests that the unusually regular geometric distribution of the taxa might cause the significant, negative BDC between the two groups. If that is the case, then the negative BDC would be a statistical artifact and not an indicator of discontinuity. Though the negative BDC appears inconclusive, it seems likely that the two groups could be monobaramins, based on the significant, positive BDC. Future studies should definitely increase the sample of taxa by including outgroup taxa from other families than the Trilliaceae.

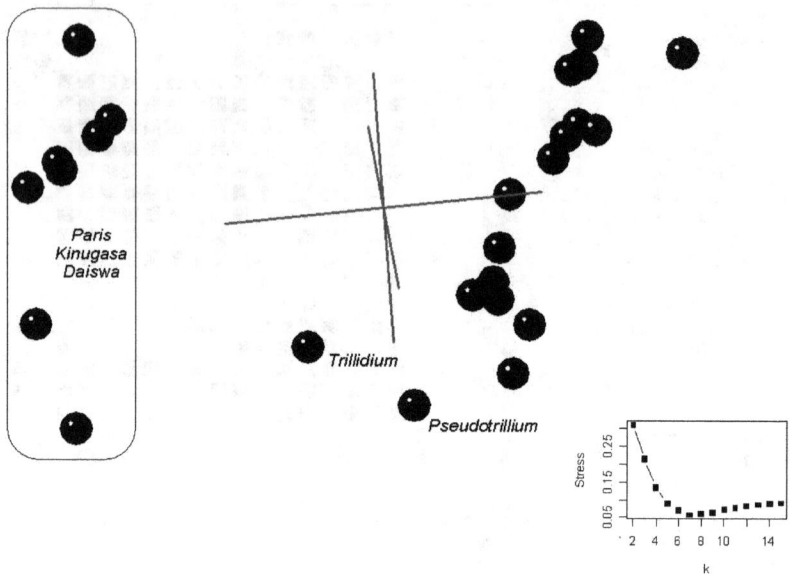

Figure 117. Three-dimensional MDS applied to Trilliaceae baraminic distances and the stress of $k$-dimensional MDS on the same baraminic distance matrix plotted as a function of the number of dimensions ($k$).

### 3.18. Cupressaceae (Pinophyta: Coniferopsida: Coniferales)

Dataset published by Gadek et al. (2000)

| | |
|---|---|
| Characters in published dataset: | 45 |
| Taxa in published dataset: | 40 |
| Character relevance cutoff: | 0.95 |
| Characters used to calculate BD: | 36 |
| Taxa used in BDC and MDS analysis: | 40 |
| Stress for 3D MDS: | 0.224 |
| $k_{min}$: | 8 |
| Median bootstrap value: | 66 |
| $F_{90}$: | 0.203 |

The cypress family Cupressaceae *sensu stricto* contains approximately 30 genera of evergreen trees, including the juniper and cypress. The Taxodiaceae (including the bald cypress *Taxodium*) are often included in the Cupressaceae based on numerous morphological and molecular phylogenetic analyses. Gadek et al.'s (2000) molecular and morphological analysis of the Cupressaceae supported the monophyly of Cupressaceae *sensu lato*. Their phylogenies showed that the subfamily Sequoioideae was generally the most basal of the Cupressaceae, with Taxodioideae (= Taxodiaceae) branching later. In no case were the Taxodioideae included among the outgroup taxa.

Gadek et al.'s (2000) morphological dataset came from phytochemical, vegetative, and reproductive characteristics. They selected 45 characters from 40 taxa. Their taxon sampling covered 38 Cupressaceae *s. l.* species from 29 genera. Three of the Cupressaceae species were members of the Taxodioideae. Also included were two outgroup species from Pinaceae and Taxaceae. For baraminic distance calculations, all taxa and 36 characters were used (omitted characters: 5, 18, 29, 38-43).

BDC results clearly separate the taxa into two groups (Figure 118). One group consists of twelve taxa from the outgroups, Taxodioideae, Sequoioideae, Taiwanioideae, Athrotaxidoideae, Cunninghamioideae, and one member of Callitroideae (*Neocallitropsis*). The second group contains the remaining taxa, including Cupressoideae and the rest of the Callitroideae. Within each group, significant, positive BDC is common and between the groups, significant, negative BDC is common. There is one taxon pair that connects the two groups by significant, positive BDC: the taxodioid *Glyptostrobus* and the cupressoid *Fokienia*, but the bootstrap value was low (58%). Overall, the bootstrap values are poor (median 66%), but high bootstrap values (>90%) are more common among the positive than the negative BDC.

The MDS results are poor (3D stress 0.224), but the distribution of taxa is generally consistent with the BDC results (Figure 119). The larger group consisting of most of the callitroids and all the cupressoids is the more closely clustered group, while the smaller group is much more diffuse. There is definitely a gap between the two groups, but the largest gap in the dataset is between the pinaceous *Picea* and the remaining taxa. In the BDC, *Picea* is positively correlated with only two taxa of the smaller group, *Amentotaxus* and *Arthrotaxis*, and negatively correlated with eight taxa of the larger group. *Glyptostrobus* is adjacent to the larger group of cupressoids and callitroids, but appears to be no closer to that group than, e.g. *Neocallitropsis*.

Taken alone, the BDC results would suggest that the cupressoids and callitroids are a holobaramin. The positive correlation within the group indicates it is a monobaramin, and the negative correlation between the two groups indicates that the larger Cupressoideae + Callitroideae group is an apobaramin. The positive BDC between *Glyptostrobus* and *Fokienia* would be interpreted as a statistical artifact. The MDS results suggest that the true discontinuity is between *Picea* and the remaining taxa, but this is poorly supported by the BDC results. Given the present set of taxa, Cupressoideae + Callitroideae can tentatively be identified as a holobaramin, pending future analyses. Additional outgroup taxa and a bigger sample of Cupressaceae *s. l.* might help to clarify the evidence of discontinuity.

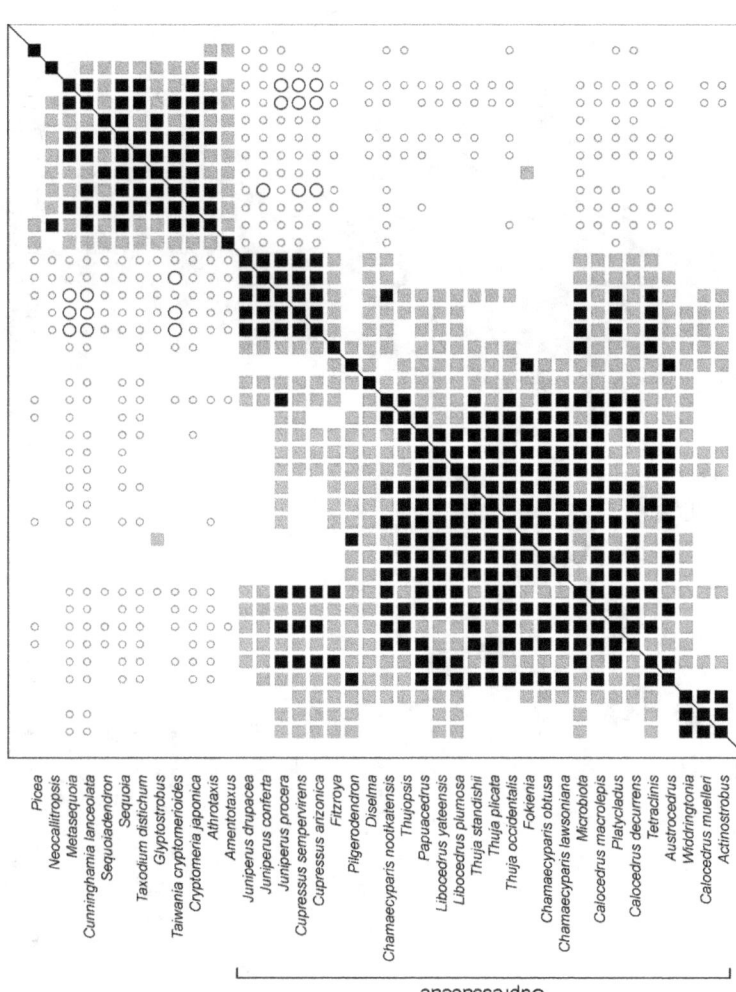

Figure 118. BDC bootstrap results for Cupressaceae, as calculated by BDISTMDS (relevance cutoff 0.95). Closed square indicate significant, positive BDC; open circles indicate significant, negative BDC. Black symbols indicate bootstrap values > 90% in a sample of 100 pseudoreplicates. Grey symbols represent bootstrap values ≤ 90%.

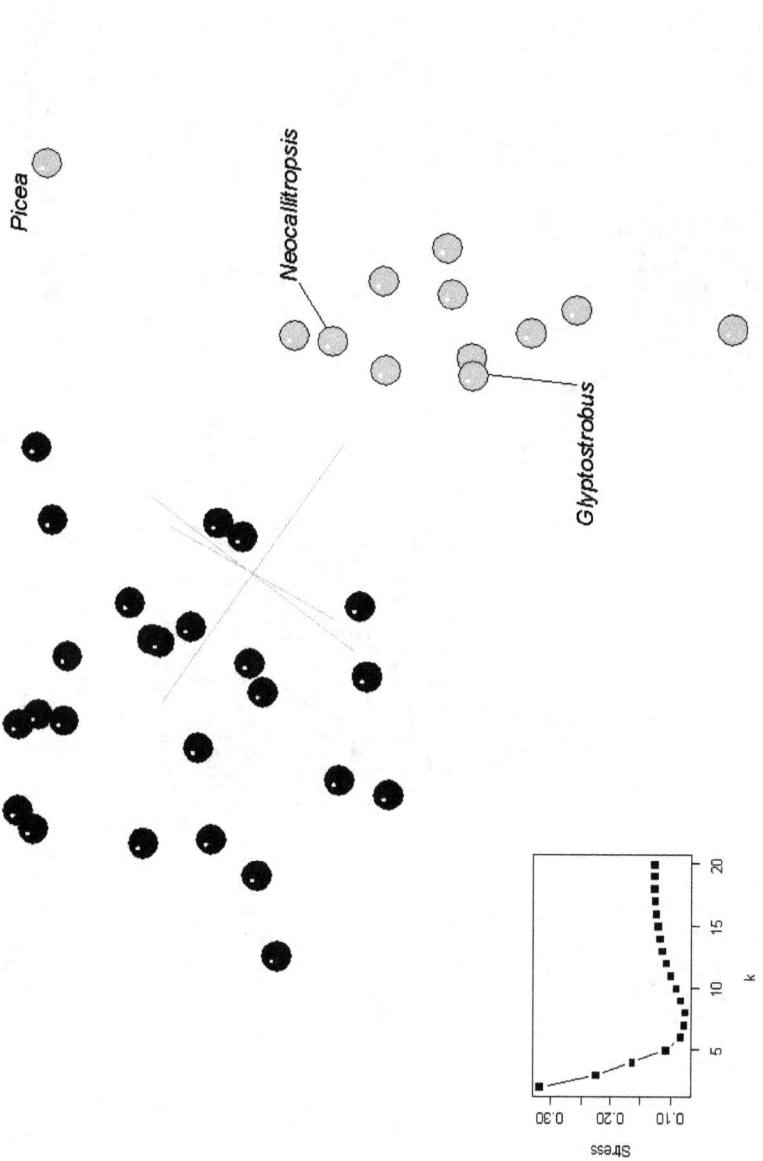

Figure 119. Three-dimensional MDS applied to Cupressaceae baraminic distances and the stress of *k*-dimensional MDS on the same baraminic distance matrix plotted as a function of the number of dimensions (*k*). Cupressaceae is shown in black, outgroup taxa in gray.

### 3.19. Podocarpaceae (Pinophyta: Coniferopsida: Coniferales)

Dataset published by Kelch (1997)

| | |
|---|---|
| Characters in published dataset: | 54 |
| Taxa in published dataset: | 25 |
| Character relevance cutoff: | 0.95 |
| Characters used to calculate BD: | 27 |
| Taxa used in BDC and MDS analysis: | 20 |
| Stress for 3D MDS: | 0.146 |
| $k_{min}$: | 5 |
| Median bootstrap value: | 69 |
| $F_{90}$: | 0.089 |

The conifer family Podocarpaceae consists of less than 200 species, mostly found in the southern hemisphere. The family also contains the only parasitic conifer, *Parasitaxus ustus*. Traditionally, podocarp species are placed in seven genera, but recent work has suggested as many as 19 genera. Kelch's (1997) morphological analysis of podocarp phylogeny supported two large clades, which he called "scale-leaved" and "tropical," but a follow-up analysis using rDNA did not support these clades (Kelch 1998). Both studies indicated that *Podocarpus* was polyphyletic or paraphyletic. Kelch's morphological dataset contained 54 characters and 25 taxa. The characters were taken from vegetative, reproductive, and embryological attributes. The taxa were entirely podocarps, no outgroups were used. The taxon sample did include five Mesozoic fossil podocarps and the extant disputed taxa *Phyllocladus* and *Saxegothaea*.

For baraminic distances, all five fossil taxa were omitted because of low taxic relevance (0.11-0.32). Even with those taxa removed, there were still only 27 characters used after filtering at a character relevance cutoff of 0.95 (omitted characters: 5-6, 12, 18-19, 22, 24, 27, 29, 36-53). The scale-leaved and tropical clades can be observed in the BDC results, although positive correlation is sparse (Figure 120). Included in the tropical clade are *Sundacarpus*, *Prumnopitys*, and *Saxegothaea*. Between the two groups, 44 of the 96 taxon pairs have significant, negative BDC, but only four of those have bootstrap values >90%. There are no taxon pairs between the two groups with significant, positive BDC. Overall the bootstrap values are exceptionally poor (median 69%), indicating that the correlations are very susceptible to character selection.

# Animal and Plant Baramins

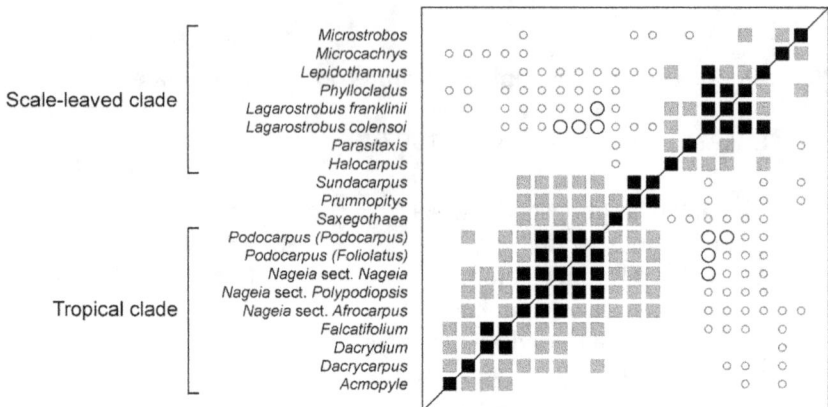

Figure 120. BDC bootstrap results for Podocarpaceae, as calculated by BDISTMDS (relevance cutoff 0.95). Closed square indicate significant, positive BDC; open circles indicate significant, negative BDC. Black symbols indicate bootstrap values >90% in a sample of 100 pseudoreplicates. Grey symbols represent bootstrap values ≤90%.

The MDS results are also somewhat poor (3D stress 0.156) and corroborate the BDC findings only in part (Figure 121). There are definitely two groups of taxa that correspond to Kelch's (1997) scale-leaved and tropical clades, but the three other podocarps (*Sundacarpus*, *Prumnopitys*, and *Saxegothaea*) are found to one side of these two clusters but not obviously a part of either one. The entire distribution of taxa is diffuse; the taxa are not closely spaced.

These results would seem to be inconclusive. The BDC analysis suggests that there is a discontinuity within the family, but the bootstrap values are exceptionally bad, indicating that the negative BDC (from which discontinuity is inferred) could merely be an artifact of the particular characters used. The distribution of taxa in MDS confirms the existence of two different groups, but three taxa (*Sundacarpus*, *Prumnopitys*, and *Saxegothaea*) appear to be intermediate between them. Given the poor quality of both the BDC and MDS results, it is best to refrain from drawing any baraminological conclusions from this analysis.

# Plants 211

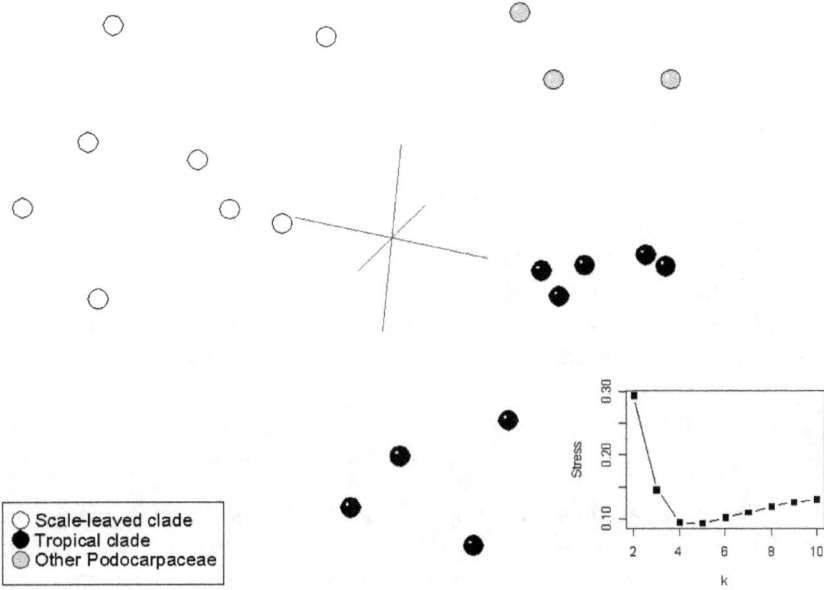

Figure 121. Three-dimensional MDS applied to Podocarpaceae baraminic distances and the stress of $k$-dimensional MDS on the same baraminic distance matrix plotted as a function of the number of dimensions ($k$). Members of the tropical clade are shown in black, of the scale-leaved clade in white, and other podocarps in gray.

### 3.20. Grammitidaceae (Polypodiophyta: Filicopsida: Filicales)

Dataset published by Ranker et al. (2004)

| | |
|---|---|
| Characters in published dataset: | 87 |
| Taxa in published dataset: | 80 |
| Character relevance cutoff: | 0.9 |
| Characters used to calculate BD: | 64 |
| Taxa used in BDC and MDS analysis: | 80 |
| Stress for 3D MDS: | 0.289 |
| $k_{min}$: | 9 |
| Median bootstrap value: | 96 |
| $F_{90}$: | 0.653 |

Grammitidaceae is a group of about 750 ferns, often placed in the family Polypodiaceae. Grammitid species are frequently epiphytic and are generally found in tropical or subtropical regions. Ranker et al. (2004) evaluated the phylogeny of 75 grammitids and five outgroups from the Polypodiaceae. They used sequence and morphological data, and found good support for the monophyly of Grammitidaceae. Their morphological dataset consisted of 87 vegetative and reproductive characters.

For baraminic distance calculations, all taxa and 64 characters were used. The BDC results show two strongly supported groups: grammitids and outgroups (Figure 122). Within the grammitids, 2709 of 2775 taxon pairs had significant, positive BDC, and 1911 of those had bootstrap values >90%. No taxon pairs involving two grammitids had significant, negative BDC. The outgroup taxa are separated from the grammitids by 351 instances of significant, negative BDC out of 375 taxon pairs. No instances of significant, positive BDC were observed between ingroup and outgroup taxa. The negative correlation between the outgroup and grammitids generally had high bootstrap values; 286 taxon pairs had bootstrap values >90%. Overall the bootstrap values were excellent (median 96%).

Surprisingly, the MDS results are exceptionally poor, with a 3D stress of 0.289, but the taxon distribution generally supports the BDC results (Figure 123). The grammitids form a tight cluster separated from the outgroup taxa. Among the outgroups, *Pecluma alfredii* approaches closest to the grammitids but is still separated from that group.

Based on these results, Grammitidaceae is a classic case of a holobaramin. The significant, positive BDC within the group implies the continuity of the species, and the significant, negative BDC with the outgroups implies discontinuity with other species. The high bootstrap values indicate that these results do not depend on any peculiar set of data but instead are robust to random sampling of the original dataset. The MDS results, though poor at three dimensions, nevertheless support the basic findings of the BDC, showing a tight cluster of grammitids, with a separate, more diffuse cluster of outgroup taxa. Taken together, these results support the holobaraminic status of Grammitidaceae.

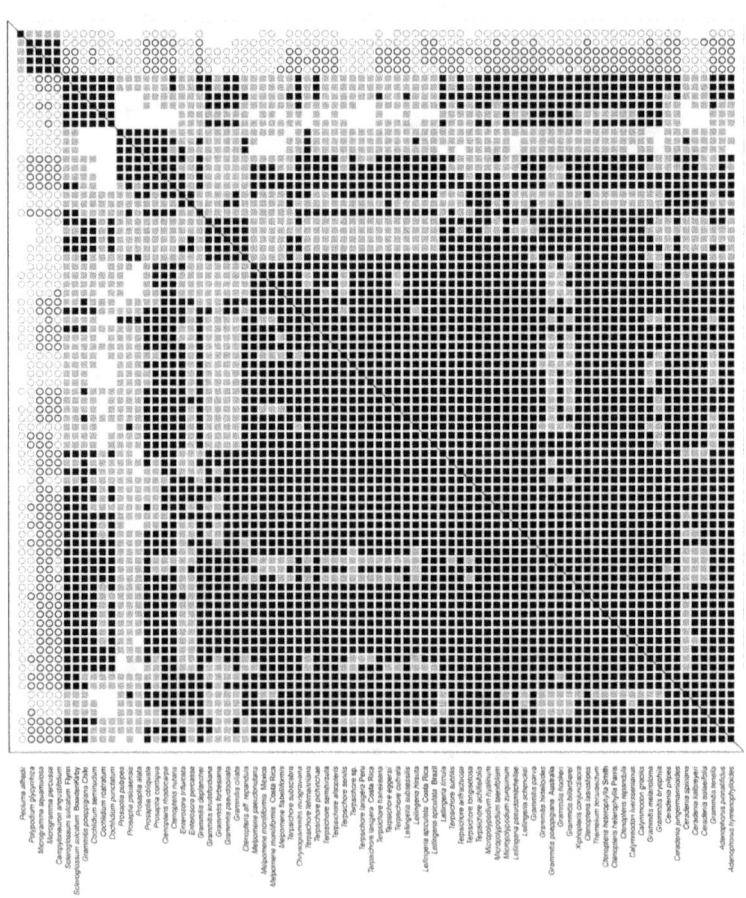

Figure 122. BDC bootstrap results for Grammitidaceae, as calculated by BDISTMDS (relevance cutoff 0.9). Closed square indicate significant, positive BDC; open circles indicate significant, negative BDC. Black symbols indicate bootstrap values > 90% in a sample of 100 pseudoreplicates. Grey symbols represent bootstrap values ≤90%.

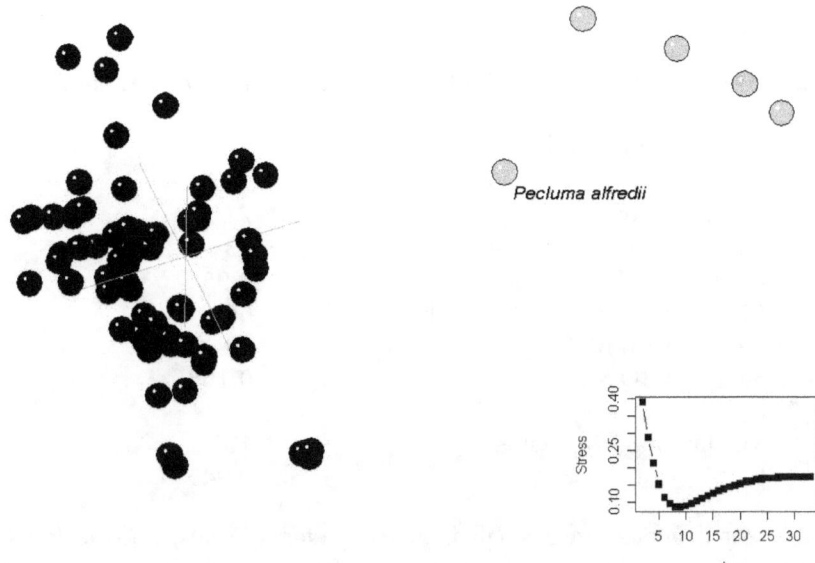

Figure 123. Three-dimensional MDS applied to Grammitidaceae baraminic distances and the stress of *k*-dimensional MDS on the same baraminic distance matrix plotted as a function of the number of dimensions (*k*). Grammitids are shown in black, outgroup taxa in gray.

### 3.21. Marsileaceae (Polypodiophyta: Filicopsida: Hydropteridales)

Dataset published by Pryer (1999)

| | |
|---|---|
| Characters in published dataset: | 71 |
| Taxa in published dataset: | 23 |
| Character relevance cutoff: | 0.95 |
| Characters used to calculate BD: | 66 |
| Taxa used in BDC and MDS analysis: | 23 |
| Stress for 3D MDS: | 0.168 |
| $k_{min}$: | 7 |
| Median bootstrap value: | 100 |
| $F_{90}$: | 0.949 |

Marsileaceae are a small family of heterosporous, aquatic ferns. There are approximately 80 species classified in 4-5 genera, the largest of which is *Marsilea*. Fossils referable to Marsileaceae are known from the Mesozoic (Lupia et al. 2000). Pryer (1999) conducted a phylogenetic study of the Marsileaceae and thirteen other fern families using 71 morphological characters. The taxon sample included five marsilaceous species from three genera and 18 outgroup taxa. The morphological dataset consisted of vegetative and reproductive characters from the sporophyte and gametophyte. Pryer (1999) found that two genera, *Salvinia* (Salviniaceae) and *Azolla* (Azollaceae), together with the fossil *Hydropteris* (Hydropteridaceae), consistently formed the sister taxa to the Marsileaceae.

For baraminic distances, all taxa and 66 of Pryer's (1999) characters were used (omitted characters: 42, 45-46, 61, 64). The BDC results reveal two well-defined groups of taxa (Figure 124). The first group consists of Marsileaceae, *Salvinia*, *Azolla*, and *Hydropteris*. All taxon pairs within this group have significant, positive BDC. The second group consists of the remaining outgroup taxa (ten families), with all taxon pairs having significant, positive BDC and bootstrap values >90%. Between the two groups, every taxon pair is negatively correlated, with all but five having bootstrap values >90%. Overall, the bootstrap values are outstanding, with 94.9% of all taxon pairs having bootstrap values >90%.

The MDS results show two groups of taxa, and the marsileaceous group is more diffuse (Figure 125). There does seem to be a space between the Marsileaceae *sensu stricto* and *Azolla*, *Salvinia*, and *Hydropteris*, but the marsilaceous taxa are also spread out. The greatest separation is between the two groups observed in the BDC results.

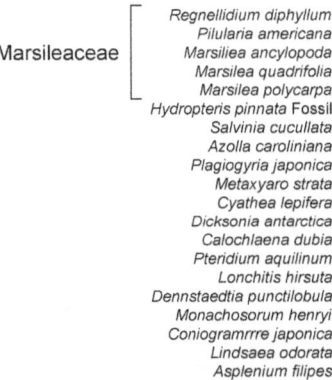
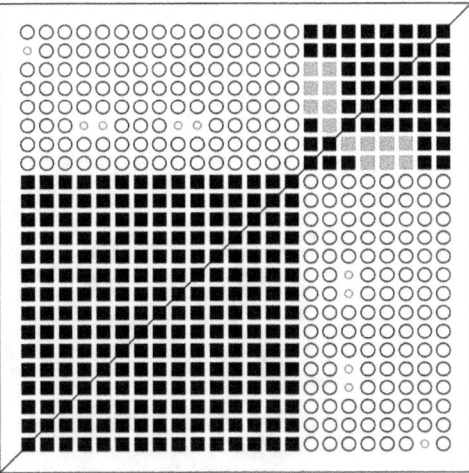

Figure 124. BDC bootstrap results for Marsileaceae, as calculated by BDISTMDS (relevance cutoff 0.95). Closed square indicate significant, positive BDC; open circles indicate significant, negative BDC. Black symbols indicate bootstrap values >90% in a sample of 100 pseudoreplicates. Grey symbols represent bootstrap values ≤90%.

Given these results, Marsileaceae, *Salvinia*, *Azolla*, and *Hydropteris* (hereafter Marsileaceae *sensu lato*) may be classified in a single holobaramin. The BDC results fully support this conclusion, and the impeccable bootstrap results indicate that the BDC pattern is robust to random character sampling. Furthermore, the MDS results, though poor (3D stress 0.168), also support the separation of Marsileaceae *s. l.* from the other ferns in the dataset. Inclusion of these additional taxa does not entail greatly expanding the marsileaceous family, since Azollaceae, Salviniaceae, and Hydropteridaceae are each monogeneric with six, ten, and one species respectively. Though future studies could expand the sample of marsileaceous taxa, it seems unlikely that the holobaramin here identified would be altered significantly.

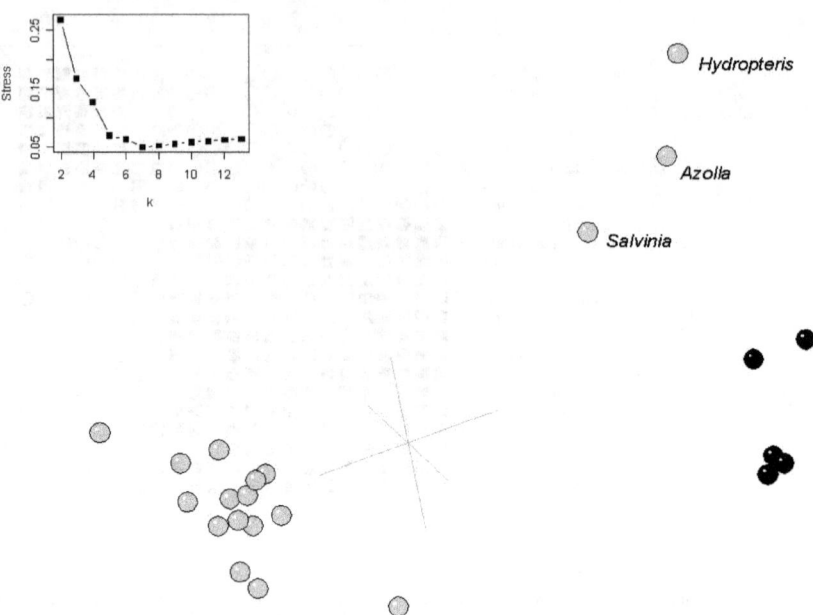

Figure 125. Three-dimensional MDS applied to Marsileaceae baraminic distances and the stress of k-dimensional MDS on the same baraminic distance matrix plotted as a function of the number of dimensions (k). Marsileaceae is shown in black, outgroup taxa in gray.

### 3.22. Bryaceae (Bryophyta: Bryopsida: Bryales)

Dataset published by Pedersen et al. (2003)

| | |
|---|---|
| Characters in published dataset: | 39 |
| Taxa in published dataset: | 53 |
| Character relevance cutoff: | 0.95 |
| Characters used to calculate BD: | 34 |
| Taxa used in BDC and MDS analysis: | 53 |
| Stress for 3D MDS: | 0.182 |
| $k_{min}$: | 6 |
| Median bootstrap value: | 68.5 |
| $F_{90}$: | 0.298 |

Bryaceae is a family of acrocarpous mosses found worldwide on all continents. There are approximately 15 genera and 500 species of moss in this family. Pedersen et al.'s (2003) phylogenetic analysis of the family included morphological and molecular data. Their morphological dataset consisted of 39 characters and 53 taxa. The characters came from vegetative and reproductive attributes. The sample of taxa included 51 Bryaceae species and two outgroup taxa representing Mniaceae (*Mnium*) and Leptostomaceae (*Leptostomum*). They found that larger genera such as *Bryum* were not monophyletic.

All taxa and 34 characters were used to calculate baraminic distances (omitted characters: 12, 17, 32-34). The BDC results showed one large group of taxa, including both outgroups, with sporadic positive BDC (Figure 126). The outgroup *Mnium* is positively correlated with 27 members of Bryaceae (particularly with species of *Bryum*) and negatively correlated with 15 Bryaceae taxa. *Leptostomum* is positively correlated with 30 Bryaceae taxa and negatively correlated with only four (all species of *Rhodobryum*). The two outgroups are negatively correlated with each other. The genus Rhodobryum, here represented by four species, forms a somewhat separate group, positively correlated with three different Bryaceae taxa but negatively correlated with 30.

The MDS results are poor (3D stress 0.182), but the results confirm the distribution of taxa in the BDC analysis (Figure 127). The Bryaceae appear as a flat cluster with the *Rhodobryum* species sticking out at a 90 degree angle. *Mnium* is adjacent to the *Rhodobryum* cluster, and *Leptostomum* is located within the main, flattened cluster of Bryaceae.

These results are somewhat difficult to evaluate from a baraminological perspective. The BDC analysis would not support an inference of discontinuity between the ingroup and outgroup taxa. Though the monogeneric Leptostomaceae has been included in Bryaceae in the past (e.g. see Andrews 1951) and could justifiably be included again according to the BDC results, the positive correlation of *Mnium* with 27 Bryaceae taxa is more puzzling. The MDS results suggest that *Mnium* is actually separated from the Bryaceae but due to its peculiar position in a corner between the main cluster and the *Rhodobryum* species, it appears positively correlated when there is no actual continuity. At best, the family Bryaceae could be classified as a monobaramin based on these results, possibly also including the Leptostomaceae.

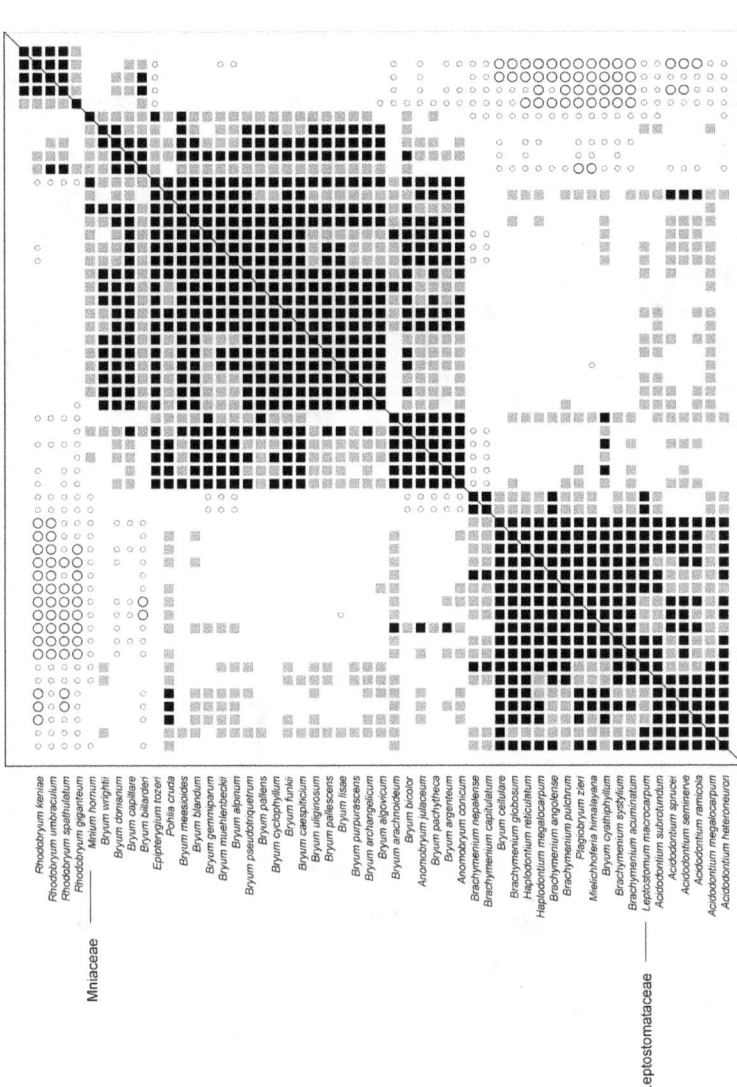

Figure 126. BDC bootstrap results for Bryaceae, as calculated by BDISTMDS (relevance cutoff 0.95). Closed square indicate significant, positive BDC; open circles indicate significant, negative BDC. Black symbols indicate bootstrap values > 90% in a sample of 100 pseudoreplicates. Grey symbols represent bootstrap values ≤ 90%.

222　Animal and Plant Baramins

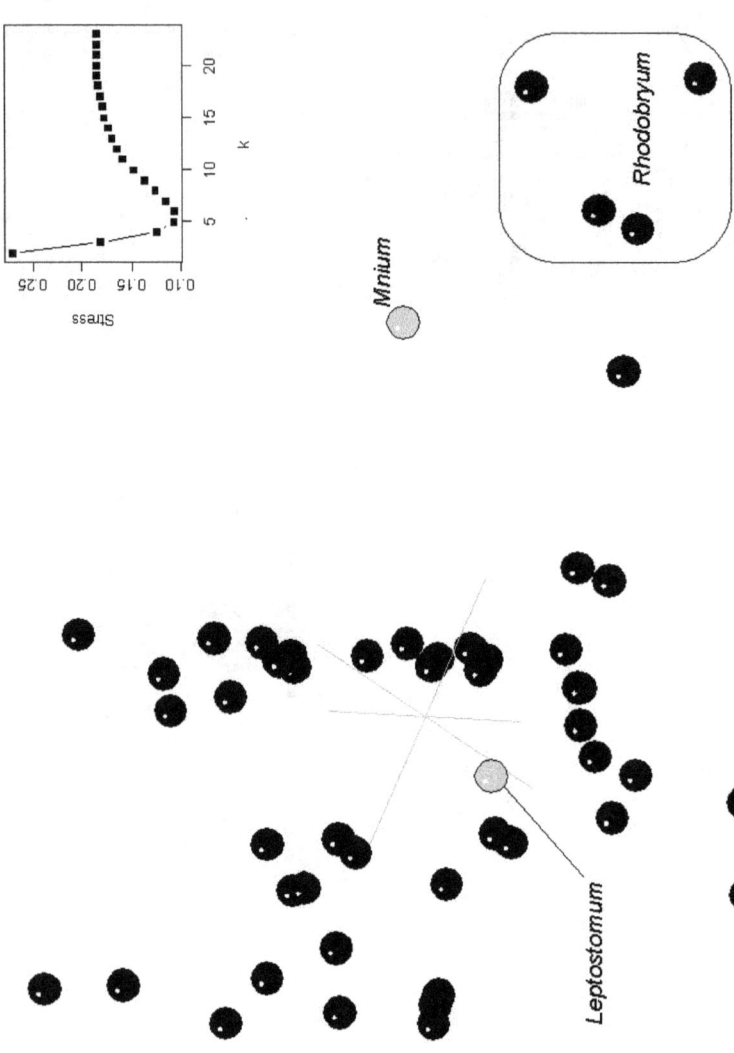

Figure 127. Three-dimensional MDS applied to Bryaceae baraminic distances and the stress of $k$-dimensional MDS on the same baraminic distance matrix plotted as a function of the number of dimensions ($k$). Bryaceae is shown in black, other taxa in gray.

# 4. Learning from the Holobaramin

## 4.1. Demography and Diversification

This work was conducted to expand the number of baraminology studies and therefore to provide a larger database of studies from which information about baramins and baraminology could be inferred. Prior to this research, Wood (2006) listed 43 monobaramins and 9 holobaramins (excluding those published only as abstracts). This present study expands that number significantly. Evidence presented here indicates an additional 14 monobaramins and 41 holobaramins, which effectively quintuples the number of identified holobaramins. Of the nine holobaramins identified prior to this study, six were reptiles (five just from the turtles), and one each came from birds (penguins), mammals (cats), and flowering plants (grasses). In the present study, holobaramins are classified as follows: seven mammals, four birds (including a confirmation of the penguin holobaramin), four fish, twelve arthropods, one annelid, ten flowering plants, one gymnosperm, and two ferns. These are the first identified holobaramins from the fish, arthropods, annelids, gymnosperms, and ferns. Together with previous studies, there has now been 57 monobaramins and 49 holobaramins identified.

Based on a very limited sample, Wood (2006) suggested that the family was the rank most closely approximating the holobaramin. Obviously, this can only be considered an approximation, since holobaramins are presumably real entities but the family is an undefined classification that often differs in membership according to the expert delineating it. The most extreme example from the present set of organisms is the stomiids, which have been placed in six families, although Fink (1985) place them in only one. Despite these caveats, the typical rank of the holobaramin can be somewhat informative, given Wood and Murray's (2003, p. 71) theological and scientific argument that the baramin should be somewhere between the order and species in the classification rank. Of the 49 holobaramins identified to date, ten are below the rank of family, representing recognized subfamilies, collections of tribes, or families missing tribes or other taxa. Seven additional holobaramins contain more than one family, and 32 correspond to a recognized family. Broken down by classification, five of eight mammal holobaramins are families, as are one of four bird holobaramins, four of six reptile holobaramins, one of four fish holobaramins, nine of twelve arthropod holobaramins, ten of eleven angiosperm holobaramins, and one of two fern holobaramins. The only annelid holobaramin is a family, but the only gymnosperm holobaramin is composed of multiple subfamilies of a larger family.

A precise accounting of all species in a holobaramin may be impossible for some groups, but an approximation of the number of species in each holobaramin can be instructive for theories of post-Flood diversification. For example, if holobaramins typically contain very few species, conventional theories of selection or drift could adequately account for their speciation. Alternatively, holobaramins with many species will likely require a much different kind of explanation for their diversification. When discussing diversification, the issue of ancestry arises, but since the refined baramin concept is based on morphology and recognition rather than ancestry (Wood et al. 2003), not all members of holobaramins need share a common ancestor. In discussing ancestry, then, the focus can be limited to holobaramins of land animals, which presumably were represented by a single pair (or seven individuals) on the Ark.

Of the 49 identified holobaramins, 27 presumably survived the Flood aboard the Ark: ten arthropods, four birds, seven mammals, and six reptiles. These holobaramins contain up to 3900 species, with a log normal distribution of species counts (Figure 128). The median number of species for these holobaramins is only 44, and only seven (26%) have more than 140 species: "other Scorpionoidea" (370 spp.), Ixodidae (650 spp.), Iguanidae (770 spp.), Pholcidae (800 spp.), Theridiidae (2000 spp.), Membracidae (3000 spp.), and Histeridae (3900 spp.). A log normal distribution could be produced by exponential speciation ($N_{species} = e^{Dt}$) with a normally distributed diversification rate ($D$).

The species counts for the 22 holobaramins that did not (or presumably did not) survive the Flood as a single pair on the Ark are also log normally distributed (Figure 128), but the magnitude of the counts are significantly higher. The median species count is 122, and eleven (50%) have species counts greater than 140, including six angiosperm holobaramins, two fish holobaramins, and one holobaramin each of annelids, ferns, and gymnosperms. The log normal distribution of these holobaramins suggests that they too experienced some kind of exponential diversification after the Flood, like the Ark-saved holobaramins. The greater number of species could be attributed either to greater intrinsic rates of natural increase (e.g., for annual plants and fish vs. terrestrial mammals) or to a greater number of Flood survivors.

The diversification rate $D$ results from at least three different factors: fecundity, environment, and theoretically an intrinsic variability.

# Learning from the Holobaramin 225

Organisms that reproduce rapidly or produce many young at one time will have more opportunities to produce new species than organisms with lower fecundity, given the same amount of time. In theory, baraminology could tie together two related observations: (1) the relationship of intrinsic rate of natural increase and body size (Fenchel 1974); (2) the observation that small-bodied mammalian groups tend to be more speciose than larger mammals (Gardezi and da Silva 1999). Since there has been a fixed time for speciation since the Flood, intrabaraminic diversification would predict the latter observation given the former.

Another secondary factor influencing diversification rate should be environment. Given the presumed harshness of the residual catastrophism following the Flood, survivability in the rapidly changing environments should be much lower than in today's more stable environments. Local extinctions were likely common, thereby retarding the net rate of diversification.

The final factor influencing the diversification rate could be an intrinsic propensity to vary. Given a nonrandom diversification mechanism that encompasses intentional design by God, it seems at least theoretically possible that not all holobaramins were created with the same potential to diversify. Alternatively, if the diversification mechanism was primarily Darwinian in nature (i.e. random variation sifted by natural selection), then the possibility of intrinsic variation seems less likely to be significant. However, since random variation is

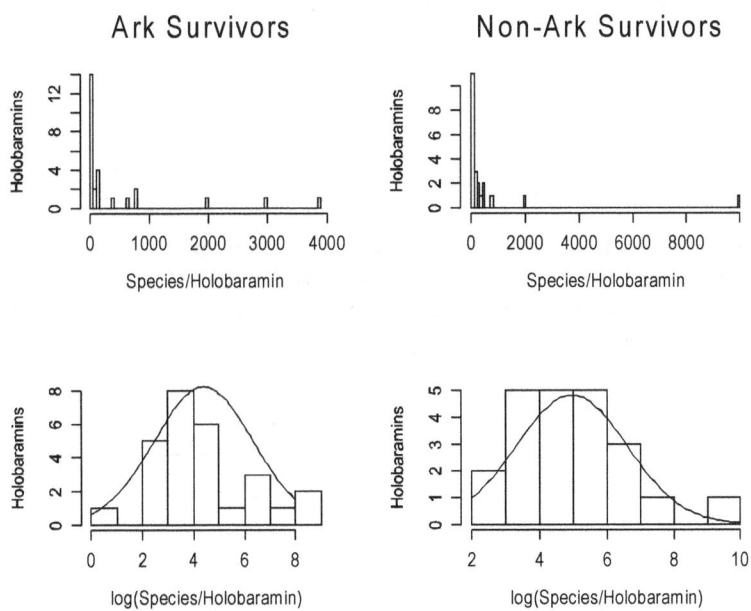

Figure 128. Log normal distribution of species counts for holobaramins preserved aboard the Ark and survivors from outside the Ark.

unlikely to produce speciation at the rates observed here, nonrandom, designed diversification is all the more likely, and intrinsic variability in the potential to diversity is also possible.

Based on scriptural and historical arguments, Wood (2003) argued that diversification occurred in a rapid burst immediately after the Flood then quickly diminished. The present results indicating an exponential speciation rate would seem to contradict that conclusion; however, if the diversification rate $D$ is largely a function of the intrinsic rate of natural increase $r$, then the intrinsic diversification rate (expressed as the number of new species born per time) could be exponentially decreasing. For example, population size at time $t$ is given by the Verhulst equation (see Murray, 1989, p. 3):

$$N_t = \frac{N_0 K e^{rt}}{K + N_0(e^{rt} - 1)}$$

[1]

where $N_0$ is the initial population size, $r$ is the intrinsic rate of natural increase, and $K$ is the carrying capacity. The fraction of individuals at any given time $t$ that will become new species would be given by

$$F_{species} = e^{-ft}$$

[2]

where $F_{species}$ is the fraction of individuals that become new species and $-f$ describes the intrinsic diversification rate.

The number of species at time $t$ is then given by the product of the fraction of new species and the population size:

$$N_{species} = \frac{N_0 K e^{(r-f)t}}{K + N_0(e^{rt} - 1)}$$

[3]

If equation [3] accurately describes the production of new species after the Flood, there are some surprising consequences (Figure 129). For all

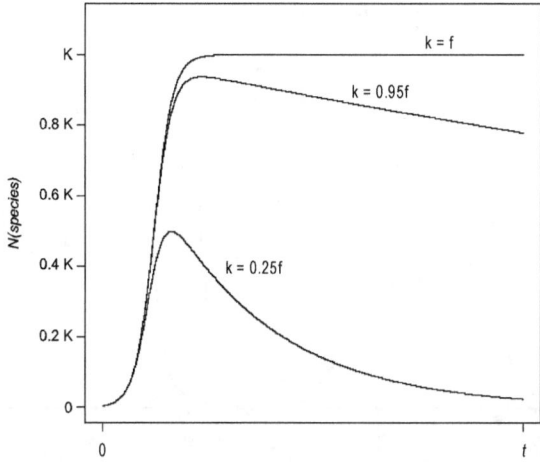

Figure 129. Species production as a function of time, using a model of speciation that assumes an exponential decay of diversification rate and population growth according to the Verhulst equation.

values of $f$, there is an initial exponential increase in species numbers, followed by a peak and then species extinctions. Looking at mammal baramins, e.g., the equids (Cavanaugh et al. 2003), where most of the species diversity is known only from the fossil record, this would seem to fit well.

Equation [3] is also unrealistic in several important ways. When $f$ is much smaller than $r$, the number of species approaches the carrying capacity $K$ before beginning a very gradual decline. In other words, there are as many species as there are individuals. For realistic situations where $N_{species} \ll K$ and $f$ is a larger fraction of $r$, the initial exponential increase in species number still occurs, but the extinction rate following it is much sharper, leading to the extinction of the entire holobaramin.

The solution to this apparent conundrum is recognizing that the carrying capacity is likely changing dramatically after the Flood, as the result of two factors: the stabilizing of the physical environment and the production of new species adapted to new niches. When $K$ is modeled as a sigmoidally increasing factor with a rate of increase $k$ (modeled after the Verhulst equation), the species growth curves become more realistic (Figure 130). In this case, when the carrying capacity grows at the same rate as the number of species ($k = f$), the number of species increases sigmoidally to the initial carrying capacity $K_0$. If the intrinsic diversification rate $f$ outstrips the growth in carrying capacity $k$ by too much (e.g. $k = 0.25\,f$), then species rapidly emerge, go extinct, and the holobaramin is wiped out.

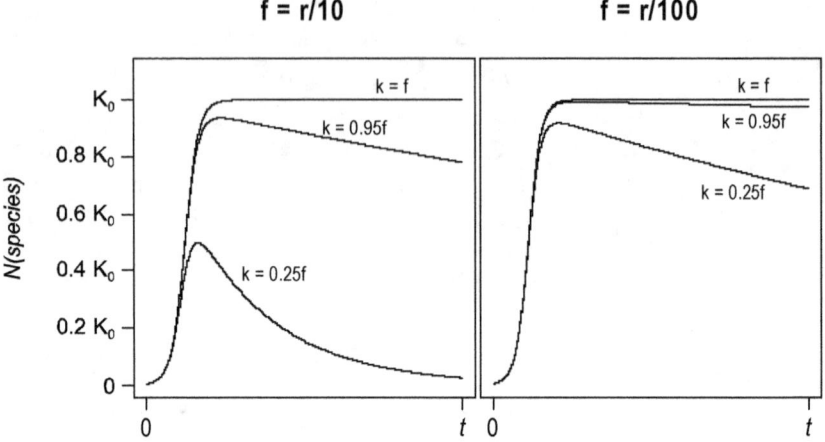

Figure 130. Species production as a function of time, using a model of speciation that assumes an exponential decay of diversification rate, population growth according to the Verhulst equation, and expansion of the carrying capacity modeled after the Verhulst equation.

### 4.2. Identifying Holobaramins

Given this large sample of BDC and MDS studies, we can now begin to address the difficult question of character selection. Put simply: What kind of characters should we use to identify holobaramins? There is very little guidance in the kind of characters to use. Wood (2002) recommended against using molecular sequence characters, and Wood et al. (2003) suggested using a "holistic" set of characters, but "holistic" was left largely undefined. Williams (2004) pointed out that merely increasing the number of characters will not necessarily change the results, and Wood (2006) suggested that further research was necessary to clarify this issue.

In the present sample of 63 datasets, holobaramins could be identified in 36, monobaramins only in thirteen, and fourteen were indeterminate. Looking at the number of characters used to calculate baraminic distances, there seems to be little pattern. Holobaramins could be identified from datasets using as little as 19 characters and as many as 305, with a mean of 70.25 characters. Monobaramins were identified from datasets of 31-252 characters (mean 80.31), and indeterminate datasets had 26-114 characters (mean 54.79). The indeterminate datasets appear on average smaller than the others, but the difference is not statistically significant ($p = 0.56$, Wilcox rank sum test). Williams (2004) is correct that merely adding characters does not improve the ability to recognize a holobaramin.

A dataset's "holism" should be simple to qualify but difficult to quantify. For example, a dataset of just dental characters would be less holistic than a dataset of craniodental, postcranial, soft tissue, and natural history characters. One extremely crude way of evaluating holism would be to count the number of common categories from which the characters are derived. For example, a dataset of just floral characters would get a score of 1, while a dataset of floral, vegetative, phytochemical, and natural history characters would get a 4. For vertebrates, dental characters would have a holism score of 1, and dental, cranial, and postcranial characters would have a score of 3. This crude estimate of complexity revealed no pattern. Holobaramin datasets had an average holism score of 2.44, and indeterminate datasets 2.35. The difference is not statistically significant ($p = 0.84$, Wilcox rank sum test).

Another way of estimating the complexity of the dataset is the stress for three dimensional MDS. High stress indicates baraminic distances (and therefore character combinations) that do not partition the taxa into well-defined groups. Stress for holobaramin datasets ranged from 0.037 to 0.305, with a mean of 0.169, and for indeterminate datasets, stress was 0.025-0.236 (mean 0.157). The stress for the indeterminate datasets is slightly lower, but this is not statistically significant ($p = 0.85$, Wilcox rank sum test).

Robinson and Cavanaugh (1998a) also recommended using a character diversity measurement to estimate the partitioning power of each individual character. Extreme diversity is bad. Diversity that is too high indicates a character where most taxa have unique character states, but low diversity implies most taxa have the same character state. The median character diversity for the holobaramin datasets was 0.178-0.645, and for indeterminate datasets 0.218-0.604. Again, the difference was not statistically significant ($p = 0.661$, Wilcox rank sum test).

Other measures of dataset quality showed differences but were also statistically insignificant. The average number of characters deleted due to low relevance was 17.36 for holobaramin datasets and 26.14 for indeterminate datasets, but this was not statistically significant ($p = 0.198$, Wilcox rank sum test). The average fraction of taxon pairs with bootstrap values $>90\%$ ($f_{90}$) was 0.443 for holobaramin datasets and 0.360 for indeterminate datasets, but again this was not significant ($p = 0.191$, Wilcox rank sum test).

Character selection will obviously be a very important area of future research in statistical baraminology, but at present the kind of characters that make a good baraminology dataset is unclear. A large number of characters is evidently not necessary for identifying a holobaramin. Nor is sampling of characters from a larger number of categories, although this might be significant, if more rigorous means of quantifying holism could be devised. The present estimate of "holism" is quite crude, and it is not surprising that it revealed no significant differences between indeterminate and holobaramin datasets. The differences seen in $f_{90}$ and the number of omitted characters due to the character relevance cutoff are also not significant. Perhaps with more datasets, these differences might turn out to be statistically significant.

### 4.3. Future Studies

The first and foremost area of future studies should be confirming the holobaramins we have and expanding the number of identified holobaramins; in short, conducting more baraminology studies. Of the 41 holobaramins identified from 36 datasets, eleven holobaramins from nine datasets were identified only tentatively. These holobaramins should be revisited with datasets containing additional taxa (to better represent the diversity of the holobaramin) or additional characters (to attempt to clarify ambiguous MDS or BDC analyses). Further, many of these large holobaramins should be examined for evidence of continuity among the species. For example, identifying a holobaramin from genera is only valid if the genera are themselves monobaramins. Confirming the continuity of these potential monobaramins will improve the confidence we can have in the holobaramin identification.

The second almost equally important issue that needs to be addressed is character selection. The superficial examination done here can be done in much greater detail. Additional analyses of other datasets can also aid by increasing the sample size, potentially allowing certain differences observed here to be statistically significant.

This large set of holobaramins can now also allow for a re-examination of the use of molecular sequences in identifying discontinuity. Many of the datasets examined here were part of a combined molecular and morphological phylogenetic study. Since these sequences are publicly available in GenBank, it is possible that examination of them will reveal differences between holobaramins.

In the larger world of creation biology, the mechanism of diversification remains mysterious. What cellular or genetic mechanism causes species to be produced so rapidly? What kind of actual speciation rates occurred? Can we infer this from identified holobaramins? A related issue is the mutational differences observed between modern cobaraminic species, as evidenced by substitutions and insertions/deletions in sequence alignments. Do these type of mutations occur constantly through time and randomly accumulate while the baramin diversifies? Or are there periods of heightened mutational occurrence, and if so, are these mutational bursts associated somehow with the generation of new species?

What implications do these baramins have for studies of natural evil? The present database of holobaramins contains many interesting test cases. The dietary habits of the phyllostomid bats, ranging from frugivory to sanguivory to insectivory, should make a promising test case to understand the origin of carnivory from herbivory after the Fall. The variation in venom toxicity in the widow spiders Theridiidae would also be an interesting test case for understanding the origins of deadly poisons. The ixodid ticks would also be an interesting group to evaluate due to their invariably parasitic lifestyle.

The other major areas of creationist biology, design and biogeography, have hardly been touched on in this text, but this larger set of holobaramins will influence these areas as well. The dispersal of species after the Flood is an important topic, for holobaramins coming off the Ark as well as holobaramins recovering from different centers around the world. How can the distribution of modern cobaraminic species be explained?

The issue of design raises perhaps the most exciting future prospect of all. Now that a good sample of holobaramins have been identified, we can develop a better understanding of superbaraminic classification, called colloquially the "pattern of life." For example, why are there such things as vertebrates? "Vertebrate" is not a holobaramin. It also cuts across the categories mentioned in the creation week (there are vertebrate flying things, swimming things, beasts of the field, and creeping things). What exactly is a vertebrate? Since this group encompasses more than one holobaramin, we can affirm that the similarity between vertebrates is due to design. Consequently, these larger categories are likely part of God's revelation of Himself in creation (Ps. 19; Rom. 1). If so, then it may be time to explore the interaction of revelation and perception using the cognitum concept of Sanders and Wise (2003). How does God reveal Himself in biology, and how do we His creations perceive Him? That is a topic worthy of great study.

# Appendix. A List of Identified Baramins
July 2008

Key:
*MB* = monobaramin (57)
*HB* = holobaramin (49)

**Kingdom Animalia**
  **Phylum Chordata**
    **Subphylum Vertebrata**
      **Class Mammalia**
        **Order Primates**
          Family Hominidae
            Subfamily Homininae *MB?* (Hartwig-Scherer 1998)
            Subfamily Australopithecinae *MB?* (Hartwig-Scherer 1998)
            Subfamily Gorillinae *MB?* (Hartwig-Scherer 1998)
            Subfamily Ponginae *MB?* (Hartwig-Scherer 1998)
          Family Cercopithecidae *MB* (Hartwig-Scherer 1993)
          Family Galagonidae *HB* (p. 5)

        **Order Xenarthra**
          Suborder Cingulata *HB?* (p. 8)

        **Order Lagomorpha**
          Family Leporidae *HB* (p. 11)

        **Order Carnivora**
          Family Canidae *MB* (Siegler 1974; Crompton 1993)
          Family Felidae *HB* (p. 14; Robinson and Cavanaugh 1998b)
          Family Ursidae *MB* (Tyler 1997)
          Family Viverridae *MB?* (p. 20)

        **Order Pinnipedia**
          Family Phocidae *HB* (p. 24)

        **Order Lipotyphla** (Erinaceomorpha)
          Family Erinaceidae
            Subfamily Erinaceinae *HB* (p. 28)
          Family Talpidae ? (p. 32)
          Family Tenrecidae ? (p. 35)

Order Chiroptera
  Family Mormoopidae *HB* (p. 39)
  Family Phyllostomidae ? (p. 42)

Order Artiodactyla
  Family Hippopotamidae (+ Anthracotheres) *HB* (p. 46)
  Family Camelidae *MB* (Wood et al. 1999)

Order Perissodactyla
  Family Brontotheriidae *MB* (p. 50)
  Family Equidae *MB* (Stein-Cadenbach 1993; Cavanaugh et al. 2003)
  Family Rhinocerotidae ? (p. 53)

Class Aves
  Order Pelecaniformes
    Family Sulidae ? (Wood 2005a)
    Family Phalacrocoracidae ? (Wood 2005a)

  Order Anseriformes
    Family Anatidae *MB* (p. 57; Scherer 1993a)
    Family Anhimidae *MB?* (Scherer 1993a)

  Order Ciconiiformes
    Family Ardeidae ? (p. 51)
  Order Falconiformes
    Family Accipitridae
      Subfamily Accipitrinae *MB?* (Zimbelmann 1993)
      Subfamily Aegypiinae *MB?* (Zimbelmann 1993)
      Subfamily Buteoninae *MB?* (Zimbelmann 1993)
    Family Cathartidae *MB* (Scherer 1993b)

    Family Falconidae
      Subfamily Falconinae *HB?* (p. 65)
      Subfamily Polyborinae *HB?* (p. 65)

  Order Galliformes
    Family Cracidae *MB?* (Klemm 1993)
    Family Megapodiidae *MB* (Klemm 1993)
    Family Phasianidae *MB* (Klemm 1993)

  Order Charadriiformes
    Family Alcidae (- Fraterculini) *HB?* (p. 68)

  Order Columbiformes
    Family Columbidae *MB?* (More 1998)

Classification 235

Order Psittaciformes
  Family Psittacidae *MB* (Jones and Mackay 1981)

Order Passeriformes
  Family Emberizidae
    Subfamily Geospizinae *MB* (**Lammerts 1966; Wood 2005a**)
  Family Estrildidae *MB* (Fehrer 1993)
  Family Fringillidae
    Subfamily Carduelinae *MB* (p. 71?; Fehrer 1993)
    Subfamily Drepanidinae *MB*? (p. 71)
  Family Pipridae *MB* (p. 71)

Order Sphenisciformes
  Family Spheniscidae *HB* (p. 81; Wood 2005a)

Class Reptilia
  Order Squamata
    Suborder Sauria
      Family Iguanidae *HB* (Wood 2005a)
        Subfamily Tropidurinae *MB* (Wood 2005a)

      Family Pygopodidae ? (p. 85)

    Suborder Serpentes
      Family Boidae
        Genus *Antaresia MB* (**Hennigan 2005**)

      Family Colubridae
        Genus *Diadophis MB* (**Hennigan 2005**)
        Genus *Nerodia MB* (**Hennigan 2005**)
        Genera *Pantherophis, Lampropeltis, Pituophis MB* (**Hennigan 2005**)
        Genus *Thamnophis MB* (**Hennigan 2005**)

      Family Pythonidae
        Genera *Morelia* & *Liasis MB* (**Hennigan 2005**)
        Genus *Python MB* (**Hennigan 2005**)

      Family Viperidae
        Genus *Agkistrodon MB* (**Hennigan 2005**)
        Genus *Bitis MB* (**Hennigan 2005**)
        Genera *Crotalus* & *Sistrurus MB* (**Hennigan 2005**)

  Order Testudines
    Family Proganochelidae *HB* (Wood 2005a)

    Suborder Pleurodira
      Family Chelidae *HB* (Wood 2005a)

Family Pelomedusidae *HB* (Wood 2005a)

Suborder Cryptodira
  Superfamilies Chelonioidea & Testudinoidea *HB* (Wood 2005a)
  Family Cheloniidae *MB* (Robinson 1997)
  Family Testudinidae
    Genus *Gopherus* *MB* (Robinson 1997)
  Superfamily Trionychoidea *HB* (Wood 2005a)

Class Amphibia
  Order Caudata
    Family Salamandridae *MB?* (p. 88)

Class Osteichthyes
  Order Gadiformes
    Family Gadidae *sensu stricto* *HB* (p. 91)

  Order Scorpaeniformes
    Family Liparidae *HB* (p. 95)

  Order Gasterosteiformes
    Family Gasterosteidae ? (p. 99)

  Order Stomiiformes
    Family Stomiidae
      Stomiid holobaramin 1 *HB?* (p. 102)
      Stomiid holobaramin 2 *HB* (p. 102)

Phylum Arthropoda
  Subphylum Crustacea
    Class Malacostraca
      Order Amphipoda
        Family Epimeriidae *HB* (p. 105)
        Family Iphimediidae *HB* (p. 102)

  Subphylum Chelicerata
    Class Arachnida
      Order Araneae
        Family Pholcidae *HB* (p. 105)
        Family Theridiidae *HB* (p. 112)

      Order Astigmata
        Family Sarcoptidae ? (p. 116)

      Order Ixodida
        Family Ixodidae *HB* (p. 120)

Order Opiliones
Family Sironidae *HB* (p. 124)

Order Scorpiones
Family Bothriuridae *HB* (p. 127)
Other Scorpionoidea (Ischnuridae, Diplocentridae, Hemiscorpiidae, Heteroscorpionidae, Urodacidae, and Scorpionidae) *HB* (p. 127)

Subphylum Uniramia
Class Insecta
Order Coleoptera
Family Curculionidae ? (Wood 2005a)
Family Histeridae *HB* (p. 131)

Order Diptera
Family Coelopidae *HB* (p. 134)

Order Hemiptera
Family Lophopidae (- *Silvana*, *Histicus*) *HB* (p. 137)
Family Membracidae *sensu stricto* (**Darninae, Membracinae, and Smiliinae**) *HB?* (p. 141)

Phylum Annelida
Class Polychaeta
Order Phyllodocida
Family Phyllodocidae *HB?* (p. 146)

Phylum Platyhelminthes
Class Trematoda
Order Strigeidida
Family Schistosomatidae
Genus *Schistosoma MB?* (Mace et al. 2003)

Kingdom Plantae
Division Magnoliophyta
Class Magnoliopsida
Subclass I Magnoliidae
Order III Piperales
Family Saururaceae
Genus *Saururus MB* (p. 149)
Genus *Gymnotheca MB* (p. 149)

Order IV Aristolochiales
Family Aristolochiaceae *HB* (p. 152)

Order VI Nymphaeales
Family Nymphaeaceae *HB?* (p. 155)

Subclass IV Dilleniidae
  Order VIII Capparidales
    Family Moringaceae ? (p. 158)

Subclass V Rosidae
  Order I Rosales
    Family Alseuosmiaceae *HB* (p. 161)
    Family Cunoniaceae *HB* (p. 164)
    Family Rosaceae
      Subfamily Maloideae *MB* (Kutzelnigg 1993b)
      Subfamily Potentilloideae
        Tribe Geeae *MB* (Junker 1993a)

  Order II Fabales
    Family Fabaceae
      Tribe Robinieae *MB?* (p. 168)

  Order IX Santalales
    "Family" Olacaceae ? (p. 171)

  Order XI Celastrales
    Family Celastraceae ? (p. 175)

Subclass VI Asteridae
  Order VIII Rubiales
    Family Rubiaceae *MB* (p. 179)

  Order XI Asterales
    Family Asteraceae
      Tribe Astereae *MB* (Wood 2005a)
      Tribes Heliantheae & Helenieae *MB* (Cavanaugh and Wood 2002)
        Subtribe Flaveriinae *MB* (Wood and Cavanaugh 2001)

Class Liliopsida
  Subclass I Alismatidae
    Order III Najadales
      Family Zosteraceae ? (p. 182)

  Subclass II Arecidae
    Order IV Arales
      Family Lemnaceae *HB* (p. 185)
      Family Araceae *HB?* (p. 185)

  Subclass III Commelinidae
    Order I Commelinales
      Family Commelinaceae *HB?* (p. 188)
      Family Rapateaceae + Mayacaceae *HB* (p. 192)

Order V Cyperales
  Family Poaceae *HB* (Wood 2002, 2005b)
    Tribe Triticeae *MB* (**Junker 1993b**)

Subclass V Liliidae
  Order I Liliales
    Family Alstroemeriaceae *HB* (p. 195)
    Family Pontederiaceae *HB?* (p. 199)
    Family Trilliaceae
      Genera *Trillium, Pseudotrillium, Trillidium* MB (**p. 202**)
      Genera *Paris, Kunigasa, Daiswa* MB (**p. 202**)

Division Pinophyta
  Class Coniferopsida
    Order Coniferales
      Family Cupressaceae
        Subfamilies Cupressoideae + Callitroideae *HB?* (p. 205)
      Family Podocarpaceae ? (p. 209)

Division Polypodiophyta
  Class Filicopsida
    Order Filicales
      Family Aspleniaceae *MB* (**Kutzelnigg 1993a**)
      Family Grammitidaceae *HB* (p. 212)

    Order Hydropteridales
      Family Marsileaceae *sensu lato* (**including** *Azolla, Salvinia,* **and** *Hydropteris*) *HB* (**p. 216**)

Division Bryophyta
  Class Bryopsida
    Order Bryales
      Family Bryaceae *MB?* (p. 219)

    Order Funariales
      Family Funariaceae *MB* (**Adler 1993**)

# References

Aagesen, L. and A.M. Sanso. 2003. The phylogeny of the Alstroemeriaceae, based on morphology, rps16 intron, and rbcL sequence data. *Systematic Botany* 28(1):47-69.

Adler M. 1993. Merkmalsausbildung und Hybridisierung bei Funariaceen (Bryophyta, Musci). In: Scherer, S., ed. *Typen des Lebens*. Pascal-Verlag, Berlin, pp. 67-70.

Agnarsson, I. 2004. Morphological phylogeny of cobweb spiders and their relatives (Araneae, Araneoidea, Theridiidae). *Zoological Journal of the Linnean Society* 141:447-626.

Andrews, A.L. 1951. Taxonomic notes X. The family Leptostomaceae. *The Bryologist* 54(4):217-223.

Arnedo, M.A., J. Coddington, I. Agnarsson, and R.G. Gillespie. 2004. From a comb to a tree: phylogenetic relationships of the comb-footed spiders (Araneae, Theridiidae) inferred from nuclear and mitochondrial genes. *Molecular Phylogenetics and Evolution* 31:225-245.

Asher, R.J. 1999. A Morphological Basis for Assessing the Phylogeny of the "Tenrecoidea" (Mammalia, Lipotyphla). *Cladistics* 15(3):231-252.

Baker, A.J., S.L. Pereira, and T.A. Paton. 2007. Phylogenetic relationships and divergence times of Charadriiformes genera: multigene evidence for the Cretaceous origin of at least 14 clades of shorebirds. *Biology Letters* 3(2):205-209.

Bininda-Emonds, O.R.P. and A.P. Russell. 1996. A morphological perspective on the phylogenetic relationships of the extant phocid seals (Mammalia: Carnivora: Phocidae). *Bonner Zoologische Monographien* 41:1-256.

Black, W.C. and J. Piesman. 1994. Phylogeny of hard- and soft-tick (Acari: Ixodida) based on mitochondrial 16S rDNA sequences. *Proceedings of the National Academy of Sciences USA* 91:10034-10038.

Boisserie, J.R., F. Lihoreau, and M. Brunet. 2005. The position of Hippopotamidae within Cetartiodactyla. *Proceedings of the National Academy of Sciences USA* 102(5):1537-1541.

Bradford, J.C. and R.W. Barnes. 2001. Phylogenetics and classification of Cunoniaceae (Oxalidales) using chloroplast DNA sequences and morphology. *Systematic Botany* 26(2):354-385.

Bremer, B. 1996. Combined and separate analyses of morphological and molecular data in the plant family Rubiaceae. *Cladistics* 12(1):21-40.

Bruvo-Mađarić, B., B.A. Huber, A. Steinacher, and G. Pass. 2005. Phylogeny of pholcid spiders (Araneae: Pholcidae): Combined analysis using morphology and molecules. *Molecular Phylogenetics and Evolution*

37(3):661-673.
Caterino, M.S. and A.P. Vogler. 2002. The phylogeny of the Histeroidea (Coleoptera: Staphyliniformia). *Cladistics* 18(4):394-415.
Cavanaugh, D.P. and T.C. Wood. 2002. A baraminological analysis of the tribe Heliantheae *sensu lato* (Asteraceae) using Analysis of Pattern (ANOPA). *Occasional Papers of the BSG* 1:1-11.
Cavanaugh, D.P., T.C. Wood, and K.P. Wise. 2003. Fossil Equidae: a monobaraminic, stratomorphic series. In: Ivey, R.L., ed. *Proceedings of the Fifth International Conference on Creationism*. Creation Science Fellowship, Pittsburgh, PA, pp. 143-153.
Cerdeño, E. 1995. Cladistic analysis of the family Rhinocerotidae (Perissodactyla). *American Museum Novitates* 3143: 1-25.
Chu, P.C. 2002. A morphological test of the monophyly of the cardueline finches (Aves: Fringillidae, Carduelinae). *Cladistics* 18:279-312.
Crompton, N.E.A. 1993. A review of selected features of the family Canidae with reference to its fundamental taxonomic status. In: Scherer, S., ed. *Typen des Lebens*. Pascal-Verlag, Berlin, pp. 217-224.
Cronquist, A. 1981. *An Integrated System of Classification of Flowering Plants*. Columbia University Press, New York.
Crosby, G.T. 1972. Spread of the cattle egret in the western hemisphere. *Bird Banding* 43(3):205-212.
Cryan, J.R., B.M. Wiegmann, L.L. Deitz, C.H. Dietrich, and M.F. Whiting. 2004. Treehopper trees: Phylogeny of Membracidae (Hemiptera: Cicadomorpha: Membracoidea) based on molecules and morphology. *Systematic Entomology* 29:441-454.
de Bivort, B.L. and G. Giribet. 2004. A new genus of cyphophthalmid from the Iberian Peninsula with a phylogenetic analysis of the Sironidae (Arachnida : Opiliones : Cyphophthalmi) and a SEM database of external morphology. *Invertebrate Systematics* 18(1):7-52.
Eckenwalder, J.E. and S.C.H. Barrett. 1986. Phylogenetic systematics of Pontederiaceae. *Systematic Botany* 11(3):373-391.
Eklöf, J., F. Pleijel, and P. Sundberg. 2007. Phylogeny of benthic Phyllodocidae (Polychaeta) based on morphological and molecular data. *Molecular Phylogenetics and Evolution* 45:261-271.
Ericson, P.B.P., U.S. Johansson, and T.J. Parsons. 2000. Major divisions in oscines revealed by insertions in the nuclear gene c-*myc*: a novel gene in avian phylogenetics. *The Auk* 117(4):1069-1078.
Evans, T.M., R.B. Faden, M.G. Simpson, and K.J. Sytsma. 2000. Phylogenetic relationships in the Commelinaceae: I. A cladistic analysis of morphological data. *Systematic Botany* 25(4):668-691.
Farmer, S.B. and E.E. Schilling. 2002. Phylogenetic analyses of Trilliaceae based on morphological and molecular data. *Systematic Botany* 27(4):674-692.
Fehrer, J. 1993. Interspecies-Kreuzungen bei cardueliden Finken und Prachtfinken. In: Scherer, S., ed. *Typen des Lebens*. Pascal-Verlag, Berlin, pp. 197-215.
Fenchel, T. 1974. Intrinsic rate of natural increase: the relationship with

body size. *Oecologia* 14:317-326.
Fink, W.L. 1985. Phylogenetic Interrelationships of the Stomiid Fishes (Teleostei: Stomiiformes). *University of Michigan Museum of Zoology Miscellaneous Publications* 171:1-127.
Frair, W. 2000. Baraminology - classification of created organisms. *Creation Research Society Quarterly* 37:82-91.
Gadek, P.A., D.L. Alpers, M.M. Heslewood, and C.J. Quinn. 2000. Relationships within Cupressaceae *sensu lato*: a combined morphological and molecular approach. *American Journal of Botany* 87:1044-1057.
Gardezi, T. and J. da Silva. 1999. Diversity in relation to body size in mammals: a comparative study. *The American Naturalist* 153(1):110-123.
Gaubert, P., W.C. Wozencraft, P. Cordeiro-Estrela, and G. Veron. 2005. Mosaics of convergences and noise in morphological phylogenies: what's in a viverrid-like carnivoran? *Systematic Biology* 54(6):865-894.
Gaudin, T.J. and J.R. Wible. 2005. The phylogeny of living and extinct armadillos (Mammalia, Xenarthra, Cingulata): a craniodental analysis. In: Carrano, M.T., T.J. Gaudin, R.W. Blob, and J.R. Wible, eds. *Amniote Paleobiology: Perspectives on the Evolution of Mammals, Birds and Reptiles*. University of Chicago Press, Chicago, pp. 153-198.
Givnish, T.J., T.M. Evans, M.L. Zjhra, T.B. Patterson, P.E. Berry, and K.J. Sytsma. 2000. Molecular evolution, adaptive radiation, and geographic diversification in the amphiatlantic family Rapateaceae: evidence from ndhF sequences and morphology. *Evolution* 54(6):1915-1937.
Gould, G.C. 1995. Hedgehog phylogeny (Mammalia, Erinaceidae): the reciprocal illumination of the quick and the dead. *American Museum Novitates* 3131:1-45.
Graham, S.W., J.R. Kohn, B.R. Morton, J.E. Eckenwalder, and C.H.S. Barrett. 1998. Phylogenetic congruence and discordance among one morphological and three molecular data sets from Pontederiaceae. *Systematic Biology* 47(4):545-567.
Gray, A.P. 1958. *Bird Hybrids*. Commonwealth Agricultural Bureaux, Farnham Royal, Bucks, England.
Griffiths, C.S. 1999. Phylogeny of the Falconidae inferred from molecular and morphological data. *The Auk* 116(1):116-130.
Hartwig-Scherer, S. 1993. Hybridisierung und artbildung bei den Meerkatzenartigen (Primates, Cercopithecoidea). In: Scherer, S., ed. *Typen des Lebens*. Pascal-Verlag, Berlin, pp. 245-257.
Hartwig-Scherer, S. 1998. Apes or ancestors? Interpretations of the hominid fossil record within evolutionary and basic type biology. In: Dembski, W.A., ed. *Mere Creation*, InterVarsity Press, Downers Grove, IL, pp. 212-235.
Hennigan, T. 2005. An initial investigation into the baraminology of snakes: order - Squamata, suborder Serpentes. *Creation Research Society Quarterly* 42(3):153-160.

Howes, G.J. 1991. Biogeography of gadoid fishes. *Journal of Biogeography* 18(6):595-622.
Huber, B.A. 2000. New World pholcid spiders (Araneae: Pholcidae): a revision at generic level. *Bulletin of the American Museum of Natural History* 254:1-348.
James, H.F. 2004. The osteology and phylogeny of the Hawaiian finch radiation (Fringillidae: Drepanidini), including extinct taxa. *Zoological Journal of the Linnean Society* 141:207-255.
Jennings, W.B., E.R. Pianka, and S. Donnellan. 2003. Systematics of the lizard family Pygopodidae with implications for the diversification of Australian temperate biotas. *Systematic Biology* 52(6):757-780.
Jones, A.J. 1973. How many animals in the Ark? *Creation Research Society Quarterly* 10:102-108.
Jones, D. and J. Mackay. 1981. Parrots and Noah's Flood. *Ex Nihilo* 4(3):15-18.
Junker, R. 1993a. Die Gatungen *Geum* (Nelkenwurz), *Coluria* und *Waldsteinia* (Rosaceae, Tribus Geeae). In: Scherer, S., ed. *Typen des Lebens*. Pascal-Verlag, Berlin, pp. 95-111.
Junker, R. 1993b. Der Grundtyp der Weizenartigen (Poaceae, tribus Triticeae). In: Scherer, S., ed. *Typen des Lebens*. Pascal-Verlag, Berlin, pp. 75-93.
Kårehed, J., J. Lundberg, B. Bremer, and K. Bremer. 1999. Evolution of the Australasian families Alseuosmiaceae, Argophyllaceae, and Phellinaceae. *Systematic Botany* 24(4):660-682.
Kelch, D.G. 1997. The phylogeny of the Podocarpaceae based on morphological evidence. *Systematic Botany* 22(1):113-131.
Kelch, D.G. 1998. Phylogeny of Podocarpaceae: comparison of evidence from morphology and 18S rDNA. *American Journal of Botany* 85(7):986-996.
Kelly, L.M. and F. González. 2003. Phylogenetic relationships in Aristolochiaceae. *Systematic Botany* 28(2):236-249.
Klemm, R. 1993. Die Hühnervögel (Galliformes): Taxonomische Aspekte unter besonderer Berücksichtigung artübergreifender Kreuzungen. In: Scherer, S., ed. *Typen des Lebens*. Pascal-Verlag, Berlin, pp. 159-184.
Klicka, J., K.P. Johnson, and S.M. Lanyon. 2000. New world nine-primaried oscine relationships: constructing a mitochondrial DNA framework. *The Auk* 117(2):321-336.
Klompen, J.S.H. 1992. Phylogenetic relationships in the mite family Sarcoptidae (Acari: Astigmata). *University of Michigan Museum of Zoology Miscellaneous Publications* 180: 1-154.
Klompen, J.S.H., W.C. Black, J.E. Keirans, and D.E. Norris. 2000. Systematics and biogeography of hard ticks, a total evidence approach. *Cladistics* 16(1):79-102.
Kluge, A.G. 1976. Phylogenetic relationships in the lizard family Pygopodidae: an evaluation of theory, methods and data. *University of Michigan Museum of Zoology Miscellaneous Publications* 152: 1-72.

Knudsen, S.W., P.R. Møller, and P. Gravlund. 2007. Phylogeny of the snailfishes (Teleostei: Liparidae) based on molecular and morphological data. *Molecular Phylogenetics and Evolution* 44(2):649-666.
Ksepka, D.T., S. Bertelli, and N.P. Giannini. 2006. The phylogeny of the living and fossil Sphenisciformes (penguins). *Cladistics* 22:412-441.
Kutzelnigg, H. 1993a. Die Streifenfarngewächse (Filicatae, Aspleniaceae) im grundtypmodell. In: Scherer, S., ed. *Typen des Lebens*. Pascal-Verlag, Berlin, pp. 71-74.
Kutzelnigg, H. 1993b. Verwandtschaftliche Beziehungen zwischen den Gattungen und Arten der Kernobstgewächse (Rosaceae, Unterfamilie Maloideae). In: Scherer, S., ed. *Typen des Lebens*. Pascal-Verlag, Berlin, pp. 113-127.
Lammerts, W.E. 1966. The Galapagos Island finches. *Creation Research Society Quarterly* 3(1):73-79.
Lavin, M and M. Sousa S. 1995. Phylogenetic systematics and biogeography of the tribe Robinieae (Leguminosae). *Systematic Botany Monographs* 45:1-165.
Les, D.H., M.L. Moody, S.W.L. Jacobs, and R.J. Bayer. 2002. Systematics of seagrasses (Zosteraceae) in Australia and New Zealand. *Systematic Botany* 27(3):468-484.
Les, D.H., E.L. Schneider, D.J. Padgett, P.S. Soltis, D.E. Soltis, and M. Zanis. 1999. Phylogeny, classification and floral evolution of water lilies (Nymphaeaceae; Nymphaeales): a synthesis of non-molecular, rbcL, matK, and 18S rDNA data. *Systematic Botany* 24(1):28-46.
Linnè, C. 1806. *A General System of Nature. Animal Kingdom I. Mammalia, Birds, Amphibia, Fishes*. Lackington, Allen, and Co., London.
Livezey, B.C. 1996. A phylogenetic analysis of geese and swans (Anseriformes: Anserinae), including selected fossil species. *Systematic Biology* 45(4):415-450.
Lörz, A.N. and C. Held. 2004. A preliminary molecular and morphological phylogeny of the Antarctic Epimeriidae and Iphimediidae (Crustacea, Amphipoda). *Molecular Phylogenetics and Evolution* 31(1):4-15.
Lupia, R., H. Schneider, G.M. Moeser, K.M. Pryer, and P.R. Crane. 2000. Marsileaceae sporocarps and spores from the Late Cretaceous of Georgia, U.S.A. *International Journal of Plant Science* 161(6):975-988.
Mace, S.R., B.A. Sims, and T.C. Wood. 2003. Fellowship, creation, and schistosomes. *Impact* 357:i-iv.
Malécot, V., D.L. Nickrent, P. Baas, L. van den Oever, D. Lobreau-Callen. 2004. A morphological cladistic analysis of Olacaceae. *Systematic Botany* 29(3):569-586.
Malécot, V. and D.L. Nickrent. 2008. Molecular phylogenetic relationships of Olacaceae and related Santalales. *Systematic Botany* 33(1):97-106.
Marusik, Y.M. and D. Penney. 2004. A survey of Baltic amber Theridiidae (Araneae) inclusions, with descriptions of six new species. In: Logunov, D.V. and D. Penney, eds. *European Arachnology*. KMK Scientific Press Ltd., Moscow, pp. 201–218.

Masters, J.C. and D.J. Brothers. 2002. Lack of congruence between morphological and molecular data in reconstructing the phylogeny of the galagonidae. *American Journal of Physical Anthropology* 117(1):79-93.

Mattern, M.Y. and D.A. McLennan. 2000. Phylogeny and speciation of felids. *Cladistics* 2000 16(2):232-253.

McCarthy, E.M. 2006. *Handbook of Avian Hybrids of the World*. Oxford University Press, Oxford.

McKenna, M.C. and S.K. Bell. 1997. *Classification of Mammals Above the Species Level*. Columbia University Press, New York.

McLennan, D.A. and M.Y. Mattern. 2001. The Phylogeny of the Gasterosteidae: combining behavioral and morphological data sets. *Cladistics* 17(1):11-27.

Meier, R. and B.M. Wiegmann. 2002. A phylogenetic analysis of Coelopidae (Diptera) based on morphological and DNA sequence data. *Molecular Phylogenetics and Evolution* 25(3):393-407.

Meng, S.-W., A.W. Douglas, D.-Z. Li, Z.-D. Chen, H.-X. Liang, and J.-B. Yang. 2003. Phylogeny of Saururaceae based on morphology and five regions from three plant genomes. *Annals of the Missouri Botanical Garden* 90(4):592-602.

Mihlbachler, M.C., S.G Lucas, R.J. Emry, and B. Bayshashov. 2004. A new brontothere (Brontotheriidae, Perissodactyla, Mammalia) from the Eocene of the Ily Basin of Kazakstan and a phylogeny of Asian "horned" brontotheres. *American Museum Novitates* 3439:1-43.

More, E.R.J. 1998. The created kind - Noah's doves, ravens, and their descendents. In: Walsh, R.E., ed. *Proceedings of the Fourth International Conference on Creationism*. Creation Science Fellowship, Pittsburgh, PA, pp. 407-419.

Murray, J.D. 1989. *Mathematical Biology*. Springer-Verlag, New York.

Olson, M.E. 2002. Combining data from DNA sequences and morphology for a phylogeny of Moringaceae (Brassicales). *Systematic Botany* 27(1):55-73.

Payne, R.B. and C.J. Risley. 1976. Systematics and evolutionary relationships among the herons (Ardeidae). *University of Michigan Museum of Zoology Miscellaneous Publications* 150: 1-115.

Pedersen, N., C.J. Cox, and L. Hedenäs. 2003. Phylogeny of the moss family Bryaceae inferred from chloroplast DNA sequences and morphology. *Systematic Botany* 28(3):471-482.

Petranka, J.W. 1998. *Salamanders of the United States and Canada*. Smithsonian Institution Press, Washington, D.C.

Prendini, L. 2000. Phylogeny and classification of the superfamily Scorpionoidea Latreille 1802 (Chelicerata, Scorpiones): an exemplar approach. *Cladistics* 16(1):1-78.

Prum, R.O. 1992. Syringeal morphology, phylogeny, and evolution of the Neotropical manakins (Aves: Pipridae). *American Museum Novitates* 3043:1-65.

Pryer, K.M. 1999. Phylogeny of marsileaceous ferns and relationships of the fossil *Hydropteris pinnata* reconsidered. *International Journal of Plant Sciences* 160(5):931-954.

Ranker, T.A., A.R. Smith, B.S. Parris, J.M.O. Geiger, C.H. Haufler, S.C.K. Straub, and H. Schneider. 2004. Phylogeny and evolution of grammitid ferns (Grammitidaceae): a case of rampant morphological homoplasy. *Taxon* 53(2):415-428.

Robinson, D.A. 1997. A mitochondrial DNA analysis of the Testudine apobaramin. *Creatin Research Society Quarterly* 33:262-272.

Robinson, D.A. and D.P. Cavanaugh 1998a. A quantative approach to baraminology with examples from Catarrhine primates. *Creation Research Society Quarterly* 34:196-208.

Robinson, D.A. and D.P. Cavanaugh. 1998b. Evidence for a holobaraminic origin of the cats. *Creation Research Society Quarterly* 35:2-14.

Salles, L.O. 1992. Felid phylogenetics : extant taxa and skull morphology (Felidae, Aeluroidea). *American Museum Novitates* 3047:1-67.

Sanders, R.W. and K.P. Wise. 2003. The cognitum: a perception-dependent concept needed in baraminology. In: Ivey, R.L., ed. *Proceedings of the Fifth International Conference on Creationism*. Creation Science Fellowship, Pittsburgh, PA, pp. 445-455.

Scherer, S. 1993a. Der grundtyp der Entenartigen (Anatidae, Anseriformes): Biologische und paläontologische Streiflichter. In: Scherer, S., ed. *Typen des Lebens*. Pascal-Verlag, Berlin, pp. 131-158.

Scherer, S. 1993b. Basic types of life. In: Scherer, S., ed. *Typen des Lebens*. Pascal-Verlag, Berlin, pp. 11-30.

Schönenberger, J., E.M. Friis, M.L. Matthews, and P.K. Endress. 2001. Cunoniaceae in the Cretaceous of Europe: evidence from fossil flowers. *Annals of Botany* 88:432-437.

Siegler, H.L. 1974. The magnificence of kinds as demonstrated by the canids. *Creation Research Society Quarterly* 11:94-97.

Simmons, M.P., C.C. Clevinger, V. Savolainen, R.H. Archer, S. Mathews, and J.J. Doyle. 2001. Phylogeny of the Celastraceae inferred from phytochrome B gene sequence and morphology. *American Journal of Botany* 88:313-325.

Simmons, N.B. and T.M. Conway. 2001. Phylogenetic relationships of mormoopid bats (Chiroptera: Mormoopidae) based on morphological data. *Bulletin of the American Museum of Natural History* 258:1-100.

Soltis, D.E., P.S. Soltis, P.K. Endress, and M.W. Chase. 2005. *Phylogeny and Evolution of Angiosperms*. Sinauer, Sunderland, MA.

Soulier-Perkins, A. 1998. The Lophopidae (Hemiptera: Fulgoromorpha): Description of three new genera and key to the genera of the family. *European Journal of Entomology* 95:599-618.

Soulier-Perkins, A. 2001. The phylogeny of the Lophopidae and the impact of sexual selection and coevolutionary sexual conflict. *Cladistics* 17(1):56-78.

Stockey, R.A., G.L. Hoffman, and G.W. Rothwell. 1997. The fossil monocot *Limnobiophyllum scutatum*: resolving the phylogeny of Lemnaceae.

*American Journal of Botany* 84:355-368.

Stein-Cadenbach, H. 1993. Hybriden, Chromosomen und Artbildung bei Pferden (Equidae). In Scherer, S., ed. *Typen des Lebens.* Pascal-Verlag, Berlin, pp. 225-244.

Strauch, J.G. 1985. The Phylogeny of the Alcidae. *The Auk* 102(3):520-539.

Teletchea, F, V. Laudet, and C. Hänni. 2006. Phylogeny of the Gadidae (*sensu* Svetovidov, 1948) based on their morphology and two mitochondrial genes. *Molecular Phylogenetics and Evolution* 38(1):189-199.

Titus, T.A. and A. Larson. 1995. A molecular phylogenetic perspective on the evolutionary radiation of the salamander family Salamandridae. 44(2):125-151.

Tyler, D.J. 1997. Adaptations within the bear family: a contribution to the debate about the limits of variation. *Creation Matters* 2:1-4.

Wetterer, A.L., M.V. Rockman, and N.B. Simmons. 2000. Phylogeny of phyllostomid bats (Mammalia: Chiroptera): data from diverse morphological systems, sex chromosomes, and restriction sites. *Bulletin of the American Museum of Natural History* 248:1-200.

Whidden, H.P. 2000. Comparative myology of moles and the phylogeny of the Talpidae (Mammalia, Lipotyphla). *American Museum Novitates* 3294:1-53.

Wible, J.R. 2007. On the cranial osteology of the Lagomorpha. *Bulletin of Carnegie Museum of Natural History* 39(1):213-234.

Williams, A. 2004. Baraminology, biology and the Bible. *TJ* 18(2):53-54.

Williams, P.J. 1997. What does *min* mean? *Creation Ex Nihilo Technical Journal* 11:344-352.

Wise, K.P. 1990. Baraminology: A young-earth creation biosystematic method. In: Walsh, R.E. and C.L. Brooks, editors. *Proceedings of the Second International Conference on Creationism.* Creation Science Fellowship, Pittsburgh, pp. 345-358.

Wise, K.P. 1992. Practical Baraminology. *Creation Ex Nihilo Technical Journal* 6:122-137.

Wood, T.C. 2002. A baraminology tutorial with examples from the grasses (Poaceae). *TJ* 16(1):15-25.

Wood, T.C. 2003. Perspectives on AGEing, a young-earth creation diversification model. In: Ivey, R.L., ed. *Proceedings of the Fifth International Conference on Creationism.* Creation Science Fellowship, Pittsburgh, pp. 479-489.

Wood, T.C. 2005a. A creationist review of the history, geology, climate, and biology of the Galápagos Islands. *CORE Issues in Creation* 1:1-241.

Wood, T.C. 2005b. Visualizing baraminic distances using classical multidimensional scaling. *Origins* 57:9-29.

Wood, T.C. 2006. The current status of baraminology. *Creation Research Society Quarterly* 43(3):149-158.

Wood, T.C. 2008. Baraminic distance, bootstraps, and BDISTMDS. *Occasional Papers of the BSG* 12:1-17.

Wood, T.C. and D.P. Cavanaugh. 2001. A baraminological analysis of subtribe Flaveriinae (Asteraceae) and the origin of biological complexity. *Origins* 52:7-27.

Wood, T.C. and D.P. Cavanaugh. 2003. An evaluation of lineages and trajectories as baraminological membership criteria. *Occasional Papers of the BSG* 2:1-6.

Wood, T.C. and M.J. Murray. 2003. *Understanding the Pattern of Life*. Broadman & Holman, Nashville, TN.

Wood, T.C., P.J. Williams, K.P. Wise, and D.A. Robinson. 1999. Summaries on camel baraminology. In: Robinson, D.A. and P.J. Williams, eds. *Baraminology'99: Creation Biology for the 21$^{st}$ Century*. Baraminology Study Group, pp. 9-18.

Wood, T.C., K.P. Wise, S. Mace, K. Ingolfsland, M. Brown, and others. 2001. HybriDatabase: A computer repository of organismal hybridization data. In: Helder, M., ed. *Discontinuity: Understanding Biology in the Light of Creation*. Baraminology Study Group, p. 30.

Wood, T.C., K.P. Wise, R. Sanders, and N. Doran. 2003. A refined baramin concept. *Occasional Papers of the BSG* 3:1-14.

Yuri, T. and D.P. Mindell. 2002. Molecular phylogenetic analysis of Fringillidae, "New World nine-primaried oscines" (Aves: Passeriformes). *Molecular Phylogenetics and Evolution* 23:229-243.

Zimbelmann, F. 1993. Grundtypen bei Greifvögeln (Falconiformes). In: Scherer, S., ed. *Typen des Lebens*. Pascal-Verlag, Berlin, pp. 185-195.

# Index

Abraeinae 131
Accipiter 65
Accipitridae 65, 234
Aceratheriinae 53
Acinonyx 14, 15
Acorus 152, 185-186
Acsmithia 164
Acutalis 142
Aetalionidae 141
Aethia 68
Aethiini 68-69
Afrostyrax 176
Aktautitan 50
Alca 68
Alcidae 68-70, 234
Alcini 68-69
Alle 68
Alseuosmiaceae 161-3, 238
Alstroemeria 195-196
Alstroemeriaceae 195-198, 239
Amblyomminae 120
Ambrosina 185
Amentotaxus 206
Amphechinus 29
Amphipoda 105, 236
Anapleus 131
Anatea 112
Anatinae 57
Anaxagorea 152
Anelosimus 112
Anommochloa 113
Anser 58
Anserinae 57, 59-60
Anserini 57-58
Anthracokeryx 46
Anthracotheriidae 46-47

Anthracotherium 46
Antillotolania 141-142
Apeltes 99
Aprasia 85
Aptenodytes 81-82
Apternodus 35-36
Araceae 185-187, 238
Arales 152, 185, 238
Archaeoceti 46
Archaeopotamus 46-47
Arctocebus 5
Ardea 61
Ardeidae 61, 63-64, 234
Ardeinae 61-62
Argasidae 120-121
Argophyllaceae 161
Argophyllum 161
Argyrodinae 112
Aristolochia 152
Aristolochiaceae 152-154, 237
Arthrotaxis 206
Athrotaxidoideae 205
Aulorhynchidae 100
Averrhoa 176
Azolla 216-217, 239
Azollaceae 216-217
Bacanius 131
Balaeniceps 62
Barclaya 155-156
Barclayaceae 156
Bauera 164-165
Blarina 32
Bomarea 195-196
Botaurinae 61-62
Bothriuridae 127-128, 237
Brachyericinae 28

Brachyerix 28-29
Brachyodus 46-47
Brachyphylla 42-43
Brachyramphini 68
Brachyramphus 68
Bradypus 8
Branta 57-58
Brasenia 155
Brexia 176
Bromeliaceae 192-193
Brontopinae 50
Brontotheriidae 50-52, 234
Brosme 91
Brunellia 164-165
Brunelliaceae 164
Bryaceae 219-222, 239
Cabassous 8
Cabombaceae 155-156
Caldcluvieae 164
Callisia 188
Callitroideae 205-206, 239
Calycanthus 152
Canis 24
Caragana 168
Carduelinae 71-72, 74-75, 235-
Careproctus 95
Carollia 43
Carpodetaceae 161
Cartonema 188
Caryophyllales 171
Cathartidae 65, 234
Celastraceae 175-178, 238
Centrodontinae 141
Centrodontus 141-143
Centrotinae 141
Centruroides 127
Cephalotaceae 164
Cephalotus 164-165
Cepphini 68
Cepphus 68
Ceratophyllaceae 155
Cereopsis 57-58
Cerocida 112-113
Cerorhinca 68
Chaerilus 127
Chaetophractus 9
Chelychelynechen 57

Chileogovea 125
Chioglossa 88-89
Chlamyphorus 9
Chloranthus 152
Chloridops 73
Choeropsis 46
Chrysochloridae 35-36
Chrysospalax 35
Cicadellidae 141
Ciconiidae 61
Ciconiiformes 61-62, 234
Ciliata 91
Cinchonoideae 179
Cingulata 8, 233, 243
Cnemiornis 57
Coccothraustes 72
Cochlearius 62
Codieae 164
Coelopidae 134-237
Colchicaceae 195
Colocasia 185
Commelinaceae 188-191, 238
Condylura 32-33
Coscoroba 57
Cotingidae 78
Coula 171
Couradoa 172
Crocuta 14
Crossosoma 176
Cryptoprocta 20-21
Culaea 99-100
Cunninghamioideae 205
Cunonia 164
Cunoniaceae 164-167, 238, 247
Cunonieae 164
Cupressaceae 205-208, 239
Cupressoideae 205-206, 239
Curupira 172
Cyclopteridae 95-96
Cyclorrhynchus 68
Cygnini 57-58
Cygnus 57-58
Cylicomorpha 158-159
Cymbomorpha 142
Cynopterocoptes 116-117
Cyrtosperma 185
Cystophora 24-25

Daiswa 202, 239
Daphniphyllaceae 172
Daphniphyllum 172
Darninae 141-143, 237
Darnis 142
Dasypodidae 8-10
Deiroderes 142
Delma 85
Dendrophilinae 131
Dermacentor 120
Desmana 32-33
Desmodontinae 42
Desmodus 42
Diabolicoptinae 116
Dicerorhinus 53
Didelphis 36
Diogoa 171
Dioscorea 195
Dioscoreaceae 195
Diplocentridae 127, 237
Dipoena 112
Disporum 195
Doellotatus 9
Dolichorhinoides 50
Donaldsonia 158, 159
Drepanidinae 71-72, 76-77, 235
Drepanidini 244
Drimys 152
Echiniphimedia 105
Eggysodontinae 54
Egretta 61
Eichhornia 199
Eidmanella 113
Elaeocarpaceae 164-165
Elomeryx 46-47
Emberizinae 71
Embolotheriinae 50
Embolotherium 50
Emertonella 113
Enchelyopus 91
Endomychura 68
Engomegorna 171
Eolestes 28
Epimeriidae 105-107, 236
Epintanteoceras 50
Erignathus 24
Erinaceidae 28, 30-32, 233

Erinaceinae 28-29, 31, 233
Erythropalum 171
Estrildidae 71, 73, 235
Eucryphia 176
Eulalia 146
Eumida 146
Euphractus 9
Eupleres 20
Eupleridae 20
Euryale 155-156
Eusirus 105
Eutatus 9
Falconidae 65-67, 79, 234
Falconiformes 65, 234
Falconinae 65, 234
Felidae 2, 14, 16-20, 233
Felis 14, 15, 21
Fokienia 205-206
Fossa 20, 21
Fratercula 68
Fraterculini 68-69, 234
Fringilla 72
Fringillinae 71
Gadidae 91-94, 236
Gadinae 91-92, 94
Gaidropsarinae 91
Gaidropsarus 91
Gaindatherium 53
Galago 5-7
Galagoides 5-7
Galagonidae 5-7, 233
Galemys 32-33
Galidia 14, 20-21
Galium 179
Gasterosteus 99-100
Gavia 81
Geissoieae 164
Geochen 57
Geogalinae 35
Ghathiphimedia 105
Glossophaginae 42-43
Glyphonycteris 42
Glyptodontidae 8
Glyptostrobus 205-206
Gomphos 11
Goupia 176
Grammitidaceae 212–215, 239

Gymnotheca 149-150, 237
Hadrotarsinae 112
Haemaphysalinae 120
Haemodoraceae 188
Haemodorum 188
Heisteria 171
Hemignathus 73
Hemiscorpiidae 127-128, 237
Herpestes 20
Herpetotheres 65-66
Hesticus 137-138
Heteranthera 188
Heterocercus 78
Heterometrus 127
Heteroscorpionidae 127-128, 237
Heterozostera 182-183
Hexaprotodon 46
Hippocrateaceae 175
Hippopotamidae 46-49, 234
Hippopotamus 46
Histeroidea 241
Holocneminae 108
Houttuynia 149-150
Hyalomminae 120
Hydropteridaceae 216-217
Hydropteris 216-217, 239
Hylomyinae 28
Hylomys 29
Hyracodontidae 53-54
Hyracodontinae 54
Icaridion 134
Iphimediidae 105-107, 236
Iphimedoidea 105
Iranotherium 53
Ischnuridae 127-128, 237
Ixodes 120-121
Ixodidae 120-123, 224, 236
Ixodinae 120
Ixoroideae 179
Kenyapotamus 46-47
Kinugasa 202
Koopmania 42
Lactoris 152
Lagonosticta 71-73
Lampronycteris 42
Latrodectrinae 112
Latrodectus 112

Laurales 152-153
Lemnaceae 185-187, 238
Leopardus 14-15
Leporidae 11-13, 233
Leptostomaceae 219-220
Leptostomum 219
Lialis 85-86
Lialisinae 85-86
Libycosaurus 46-47
Liliales 152, 195, 199, 202, 239
Limnobiophyllum 185-186
Lionycteris 42
Liparidae 95-98, 236
Liparis 95
Lipotyphla 35, 233, 241, 248
Litolestes 28
Lonchorhina 42
Lophopidae 137-140, 237
Loranthaceae 171-172
Loris 5
Lota 91
Lotinae 91
Loxocephala 137-138
Lunda 68
Luzuriaga 195
Lycoderes 141
Lynx 14-15, 21
Maburea 171
Macroeuphractus 9
Macropus 36
Macrotus 39
Magnoliales 152-153
Malania 172
Marsilea 216
Marsileaceae 216-218, 239
Marwe 125
Mayaca 192-193
Mayacaceae 192-193, 238
Maytenus 176
Melamprosops 72
Membracidae 141, 143-145, 224, 237
Membracinae 141-143, 237
Mertensiella 88-89
Merycopotamus 46-47
Mesaceratherium 53
Metatitan 50

Micrastur 65-66
Microbunodon 46
Microcentrus 141-142
Microhierax 65
Micropotamogale 35
Micrutalis 142
Millettia 168-169
Mimolagus 11
Mimotona 11
Minquartia 171
Misodendronaceae 171-172
Mniaceae 219
Mnium 219-220
Molva 91
Monachinae 24-25
Monachus 24
Monoculodes 105-106
Moridae 91
Moringa 158-159
Moringaceae 158-160, 238-
Mormoopidae 39-41, 43, 234
Mormoops 39-40, 42
Mortonia 176
Mungos 20
Musonycteris 42
Mystacina 39
Mystacinidae 40
Nandinia 20-21, 36
Nelumbonaceae 155
Neocallitropsis 205-206
Neofelis 14-15
Neogoveidae 124
Nesticus 113
Neurotrichus 32-33
Ninetinae 108
Ninxiatherium 53
Noctilio 39, 42
Noctilionidae 39-40, 43
Noctilionoidea 39-43
Notophthalmus 88
Nuphar 155-156
Nycteridocoptes 117
Nycticorax 61-62
Nyctocebus 5
Nymphaceae 155-156
Nymphaea 155
Nymphaeaceae 155-157, 237

Ochanostachys 171
Ochotona 11
Ochotonidae 11
Ochropepla 142
Octoknema 171
Odobenidae 24
Odobenus 24-25
Odontosiro 124
Ogoveidae 124
Olacaceae 171-174, 238
Ondinea 155-156
Onthophilus 131-132
Ophidiocephalus 85
Opiliaceae 171-172
Orycteropus 35
Oryzorictinae 35
Otariidae 24
Otocolobus 14-15
Otolemur 5-7
Otus 65
Pakicetus 46
Palaeolagus 11-12
Palaeosyops 50
Paleuphractus 9
pampatheres 8
Panthera 14-15
Parabrontops 50
Parascalops 32
Parasiro 124
Parasitaxus 209
Parasitiformes 120
Paris 202, 239
Passer 71-73
Passeridae 71, 73
Pecluma 212
Pelecanus 65
Perodicticus 5
Perrottetia 176
Petallidae 124
Peucedraminae 71
Phalacrocoracidae 234
Pharus 113
Phellinaceae 161
Philydraceae 199
Philydrum 199, 200
Phocidae 24-27, 233
Phocinae 24-25

Phoenicopterus 62
Pholcidae 108-111, 224, 236
Pholcinae 108
Pholcommatinae 112
Phoroncidia 112
Phycinae 91
Phycis 91
Phyllocladus 209
Phyllodocidae 146-148, 237
Phyllospadix 182-183
Phyllostomidae 39-40, 42-45, 234
Phyllostominae 42-43
Physocyclus 108
Picea 206
Pinaceae 205
Pinguinus 68
Piper 152
Piperaceae 149-150
Piperales 149, 152-153, 237
Pipra 78-79
Pipridae 78-80, 235
Pistia 185-186
Pletholax 85
Pleurodeles 88-89
Poaceae 113, 172, 239
Pochazia 137-138
Podocarpaceae 209-211, 239
Podocarpus 209
Polihierax 65
Polyborinae 65, 234
Polyglypta 142
Polypodiaceae 212
Pontederiaceae 188, 199-201, 239
Potamogalinae 35
Priodontes 8-10
Prionailurus 15
Priscula 108
Proeuphractus 9
Profelis 14-15
Prolagus 11
Propalaeohoplophorus 8-9
Proterix 28
Protitan 50
Prozaedyus 9
Prumnopitys 209-210
Pseudomystides 146
Pseudotrillium 202, 239

Psittirostra 72
Ptaiochen 57
Pteronotus 39-40, 42
Ptychoramphus 68
Puma 14-15
Pungitius 99-100
Punjabitherium 53
Pygopodidae 85-87, 235
Pygopodinae 85-86
Pygopus 85
Raniceps 91
Rapateaceae 192-194, 238
Rhabdodendron 172
Rhinocerotidae 53-56, 234
Rhinocerotinae 53-54
Rhinophylla 42-43
Rhinotitan 50
Rhiphicephalinae 120
Rhodobryum 219-220
Rhombomylus 11-12
Rhynchoptidae 116-117
Ricania 137-138
Robinieae 168-170, 238
Rubiaceae 179-181, 238
Rubioideae 179
Saccopteryx 39-40
Salamandra 88-89
Salamandridae 88, 90, 236
Salvinia 216-217, 239
Salviniaceae 216-217
Santalaceae 171-172
Santalales 171-172, 238
Saotherium 46
Sarcoptes 116
Sarcoptidae 116, 118-119, 236
Sarcoptinae 116
Saruma 153
Saururaceae 149-151, 237
Saururus 149-152
Saxegothaea 209-210
Saxifragales 171-172
Scalmophorus 142-143
Scalopus 32
Scapanus 32
Scaptonyx 32-33
Schefflerodendron 168-169
Schizomerieae 164

Scleronycteris 42
Scorodocarpus 171
Scorpionidae 127-128, 237
Scorpionoidea 127, 129-130, 224, 237
Sequoioideae 205
Siamotherium 46
Sige 146
Silvanana 137-138
Siphonodon 176
Sironidae 124-126, 237
Smilacaceae 195
Smiliinae 141-143, 237
Soricidae 32
Spatholirion 188
Spheniscidae 81, 83-84, 235
Sphenisciformes 81, 235
Spinachia 99
Spintharinae 112
Spiraeanthemum 164
Spirodella 185
Spiziapteryx 65
Stegaspidinae 141-142
Stegotherium 8-10
Stenodermatinae 42-43
Stercorariidae 68
Stomiidae 102-104, 236
Streptochaeta 113
Strombosia 171
Strombosiopsis 171
Stylocellidae 124
Suidae 46
Sundacarpus 209-210
Suricata 14
Suzukielus 124
Syntelia 131-132
Synthliboramphus 68
Tadorninae 58
Taiwanioideae 205
Talpa 32
Talpidae 32-34, 233
Tamandua 8
Taricha 88
Taxaceae 205
Taxodiaceae 205
Taxodioideae 205
Taxodium 205

Tayassuidae 46
Teinocoptes 116-117
Teinocoptinae 116
Telespiza 72
Teletaceras 54
Telmatheriinae 50
Tenrecinae 35-36
Tenrecoidea 241
Tephrosia 168
Tetrastylidium 171
Tettigometra 137-138
Theridiidae 112-115, 224, 231, 236
Theridiinae 112
Threskiornithidae 61
Tigrisomatinae 61-62
Tremandraceae 164-165
Triceratella 188-189
Trichopus 152
Trilliaceae 202-204, 239
Trillidium 202, 239
Trillium 202, 239
Trilobophorus 46
Trixacarus 117
Troglosironidae 124
Trypanaeinae 131
Trypeticinae 131
Tupaiodontinae 28
Tylototriton 88-89
Tyrannidae 78
Uncia 14-15
Uria 68
Urodacidae 127-128, 237
Urophycis 91
Uropsilus 32-33
Urotrichus 32-33
Ursus 24
Uvularia 195
Uvulariaceae 195
Vampyress 42
Vassallia 8-10
Victoria 155
Viscaceae 171-172
Viscum 171
Viverridae 20-23, 233
Wisteria 168
Wolffia 185

Xenohyus 47
Ximenia 172
Zaedyus 9
Zalophus 25
Zebrilus 61-62
Zippelia 149-150
Zonerodius 61
Zostera 182-183
Zosteraceae 182-184, 238
Zosterella 182-183

# CORE Issues in Creation

Established in 2005, the *CORE Issues* monograph series presents high quality scholarly work from or related to a young-age creation perspective. This monograph series is not for the publication of scholarly critiques of alternative positions (other venues exist for that kind of publication). Rather, *CORE Issues* has been created to publish any monograph in any discipline (philosophy, theology, physics, geology, biology, archaeology, linguistics, etc., etc.) which substantially contributes to the systematic development of a positive, young-age creation model. Original monographs will thoroughly review the conventional and creationist literature on the subject, offer a constructive interpretation of the subject's data, integrate well with other disciplines as the model is constructed, and advance creation model development. Other monographs offer reprints, compendia, or translations of significant historical works that are currently unavailable. *CORE Issues* is peer-reviewed and will strive for the very highest scholarship standards. *CORE Issues* is a joint publication of the Center for Origins Research at Bryan College and Wipf & Stock Publishers.

*CORE Issues* does not publish works written only by Bryan College faculty but encourages outside submissions. Researchers may submit monograph proposals (full manuscripts are not accepted) to CORE either electronically at info@bryancore.org or by regular mail:

*CORE Issues* editor
Bryan College 7802
721 Bryan Drive
Dayton, TN 37321

**Previous Volumes in the *CORE Issues* Series**

1. A Creationist Review and Preliminary Analysis of the History, Geology, Climate, and Biology of the Galápagos Islands, by Wood (2005)
2. Johannes Buteo's The Shape and Capacity of Noah's Ark, trans. by Griffith & Monette (2008)